AF536666

Mathematics Study Resources

Volume 21

Series Editors

Kolja Knauer, Departament de Matemàtiques i Informàtica, Universitat de Barcelona, Barcelona, Spain

Elijah Liflyand, Departament of Mathematics, Bar-Ilan University, Ramat-Gan, Israel

This series comprises direct translations of successful foreign language titles, especially from the German language.

Powered by advances in automated translation, these books draw on global teaching excellence to provide students and lecturers with diverse materials for teaching and study.

Christoph Kirfel

Side Paths in the History of Mathematics

Potentials and Limits of Alternative Approaches

Springer

Christoph Kirfel
Mathematics Teacher Education
University of Bergen
Bergen, Norway

ISSN 2731-3824 ISSN 2731-3832 (electronic)
Mathematics Study Resources
ISBN 978-3-662-72917-5 ISBN 978-3-662-72918-2 (eBook)
https://doi.org/10.1007/978-3-662-72918-2

Translation from the German language edition: "Seitenwege in der Mathematikgeschichte" by Christoph Kirfel, © Der/die Herausgeber bzw. der/die Autor(en), exklusiv lizenziert an Springer-Verlag GmbH, DE, ein Teil von Springer Nature 2024. Published by Springer Berlin Heidelberg. All Rights Reserved.

This book is a translation of the original German edition "Seitenwege in der Mathematikgeschichte" by Christoph Kirfel, published by Springer-Verlag GmbH, DE in 2024. The translation was done with the help of an artificial intelligence machine translation tool. A subsequent human revision was done primarily in terms of content, so that the book will read stylistically differently from a conventional translation. Springer Nature works continuously to further the development of tools for the production of books and on the related technologies to support the authors.

© The Editor(s) (if applicable) and The Author(s), under exclusive license to Springer-Verlag GmbH, DE, part of Springer Nature 2026

This work is subject to copyright. All rights are solely and exclusively licensed by the Publisher, whether the whole or part of the material is concerned, specifically the rights of translation, reprinting, reuse of illustrations, recitation, broadcasting, reproduction on microfilms or in any other physical way, and transmission or information storage and retrieval, electronic adaptation, computer software, or by similar or dissimilar methodology now known or hereafter developed.
The use of general descriptive names, registered names, trademarks, service marks, etc. in this publication does not imply, even in the absence of a specific statement, that such names are exempt from the relevant protective laws and regulations and therefore free for general use.
The publisher, the authors and the editors are safe to assume that the advice and information in this book are believed to be true and accurate at the date of publication. Neither the publisher nor the authors or the editors give a warranty, expressed or implied, with respect to the material contained herein or for any errors or omissions that may have been made. The publisher remains neutral with regard to jurisdictional claims in published maps and institutional affiliations.

Responsible Editor: Iris Ruhmann

This Springer imprint is published by the registered company Springer-Verlag GmbH, DE, part of Springer Nature.
The registered company address is: Heidelberger Platz 3, 14197 Berlin, Germany

If disposing of this product, please recycle the paper.

Preface

Origin of the Book and Objectives

The present book project aims to revisit old methods from the history of mathematics that did not become standard methods and did not make it into textbooks. These alternative approaches were often in discussion at the time of their creation, but then lost ground in competition with what we now call standard methods, and some of them have ultimately been forgotten.

Often these methods still contain untapped potential, and it is often worthwhile to further develop these methods to see where exactly the limits of these methods lie. So it's not just about presenting historical methods that may have been forgotten, but also about their further development. Here is an example: Archimedes developed a method for calculating the area of a parabolic segment. This is an amazing achievement that anticipates the field of integral calculus. However, Archimedes' method did not develop into the standard method found in today's textbooks. There, you find the methods designed by Fermat, Newton, and Leibniz. However, there is still a residual potential in the Archimedean method that is worth discovering and developing further. Other curves, not just parabolas, can be treated similarly and it is interesting to see how far Archimedes' approach can be developed and where its limits lie.

Several articles have been published on this topic, which form the starting point for the book project.

Chapter 1 deals—as mentioned above—with the Archimedean method for determining the area of a parabolic segment and the generalization to more general polynomials. I have already described this in four articles (Kirfel 2013A, 2013B, 2014A, 2017B). Here, in Chap. 1, the Archimedean method is first presented. With minor modifications, the method can be seamlessly transferred to polynomial functions of higher degrees, as shown by some examples. However, if you exceed degree 6, the associated calculations become very tedious and you have to realize that the method also has its limits. The presentation here is much more detailed and thorough than in the mentioned articles.

Chapter 2 deals with a very old Indian method for root extraction, which was "reconstructed" by the historians Datta (1932) and Henderson (2000). It turns out that the method can be generalized easily and is suitable for all integers and fractions. With the help of this method, it is also possible to show the irrationality of roots of numbers that are not themselves perfect squares. The Indians who developed this method at the time probably did not pursue the question of the rationality of roots. However, the algorithm allows for an irrationality proof for roots, a fact that highlights the special potential of the method. There are already three articles on this topic (Kirfel 2015, 2019, 2021), which form the basis for the present book. The presentation in this chapter is, however, much more extensive and detailed than in the mentioned articles.

Chapter 3 picks up the method of Gregorius of St. Vincent (1647), who made an important contribution to the calculation of the area under the hyperbola. This method is generalized in that chapter in two different but very closely related ways, so that it ultimately covers all the functions used in school mathematics. Then, the same method is transferred to the topic of derivation, so that integral and differential calculus here get a presentation from a single source, without having to resort to the main theorem of integral and differential calculus. Six articles have already been published on this topic, four of them together with my good colleague Rainer Kaenders (University of Bonn) (Kirfel 2014B, 2018; Kaenders & Kirfel 2016, 2017, 2020A, 2020B). The chapter in this book contains more details and goes much further than the articles.

Chapter 4 revolves around two methods that Fermat developed for integration. In his writing "Sur la transformation" (Fermat 1896), he describes these methods and applies them to numerous examples. The chapter then shows how the example space can be significantly expanded, reaching function types that were still unknown to Fermat himself. There is also a precursor article on this topic (Kirfel 2017A). However, the presentation in this chapter contains a number of new aspects.

In Chap. 5, Leibniz's (2016) Resection Method is presented and clarified using a number of examples. This is followed by the fringe method, which represents the inverse of the Resection method. Here too, some examples are worked through. By reversing, the Resection Method is further developed and simultaneously illustrated geometrically. There is also a contribution (Kirfel, 2014, pp. 103–116) on this topic in the conference proceedings of the Conference for the History of Mathematics held in Leipzig, March 2023. A short version can also be found in Kirfel (2023).

Chapter 6 picks up the method of Åge Bondesen, which in turn goes back to Fermat (1896). It is a method in which Riemann rectangles are transformed into triangles, which are then "bundled" into fans, which is why we also call the method the "fan method". In this way, relationships are created between areas under different curves, which allow formulas for areas to be derived in many cases. In this way we calculate areas under complex curves by using much simpler curves. Volumes under surfaces in space can also be approached with similar considerations, as will be demonstrated at the end of this chapter. The method, which is only shown in a single example by Åge Bondesen and Fermat, can be applied to a variety of new examples, which are not exhaustively treated in this chapter. An article by me and my good colleague Helmer Aslaksen (University of Oslo) on the mentioned topic was published in early 2024 (Aslaksen & Kirfel 2024), which also adopts the method of Åge Bondesen and Fermat. However, this chapter is much richer and contains many more three-dimensional examples. A short version can be found in Kirfel (2020).

In the last chapter, we then deal with the number π. The classical approaches like Archimedes's regular polygons and the doubling of the number of corners do not form the core of the chapter, but rather the later, more algebraic, method from the 17th century, where the so-called arctangent identities became popular. With such identities and the additional use of the Gregory series for the arctangent, a variety of approximations to π can be given quickly. A method for systematically developing arctangent identities is described, and astonishing approximations for π are found. Patterns in the approximations are also discovered. However, the methods of today's high-performance athletes when calculating most decimal places of π differ significantly from this approach. Nevertheless, we can achieve interesting results and also provide insight into relationships around the number π.

In his article "History or Heritage? An Important Distinction in Mathematics and for Mathematics Education," Ivor Grattan-Guinness (2004) attempts to draw an important demarcation line in dealing with the history of mathematics. It revolves around two different approaches when it comes to the history of mathematics. In the camp of the "historians," the focus is primarily on understanding a historical fact, e.g., Pythagoras' theorem, as it was presented at the time of Pythagoras and as we know it from the Euclidean elements, as a theorem about areas of squares that can be placed on the sides of a right-angled triangle. An interpretation of the theorem as $a^2 + b^2 = c^2$, where a, b and c represent the lengths of the sides of the right-angled triangle, is an interpretation that rather belongs to the camp of the "inheritors," because here the entire algebraic heritage overshadows the original Euclidean formulation. This interpretation then has several advantages because it also includes numbers and in this case also roots. Grattan-Guinness defends the legitimacy of both approaches, but strongly advocates that we be aware of the different approaches and that we do not mix both or pretend that there are no differences. He also explicitly emphasizes the importance of both approaches in teacher education. Both approaches can enrich teaching and provide valuable opportunities for learning mathematics. In this book, we primarily align ourselves with the "heritage" point of view in the history of mathematics. We use modern notation and tools, such as the coordinate system or the concept of function, even where they had not yet been developed. This can often seem anachronistic. For example, we try to apply the method that Archimedes used to determine the area of a parabolic segment to curves of the fourth and fifth degree, although such a question would probably have been unthinkable for Archimedes, because the geometric concepts were limited to lines, areas, and three-dimensional bodies and curves of the fourth degree did not exist in such a universe. It is all the more astonishing, however, when it turns out that the Archimedean method is so ingenious and so comprehensive that it also includes curves of the fourth degree, which were unthinkable at the time of Archimedes, and that the method can solve problems that Archimedes probably could not even have formulated. At the same time, however, we are not content to simply take the "heritage" position. We want to bring a new dimension into play that extends the two poles of Grattan-Guinness. It is about a third position that sees the history of mathematics as a legacy, a mission. So, this book is not just about conveying historical methods, be it in their historical context or in today's interpretation, but also about further developing these methods. In doing so, we will sometimes also use tools that were only developed much later, but which today give us much faster access to these methods and their generalizations. The untapped residual potential of the mentioned methods is then the motivation for dealing with this material, and as the seven chapters of the book hopefully show, it is indeed worthwhile to accept the mission described above.

When dealing with mathematics, we automatically have to fall back on historically developed methods, even if this often happens unconsciously. Without thinking about it for long, we use historical notations, ideas, calculation methods, and traditions that were developed long before us. On the one hand, we can simply use these traditionally inherited achievements and use them without further thought in research and teaching. But we can also question the traditions and try to research which alternatives there were and whether these alternatives still have a right to exist today. By exploring these alternatives to further develop them, we can discover new things on the one hand, and on the other hand, we can also explore the limits of these approaches and find out how far these approaches are accessible and where it is no longer worth continuing to search on the mentioned path. This can then lead to a greater appreciation of these alternative approaches, which is also one of the aims of this book project.

An analogy can clarify the matter: Someone inherits a carpentry business from his father. Perhaps he is nostalgic and recognizes the value of old tools and methods, turning the workshop into a museum where the inherited tools and machines find their place. Perhaps he offers courses where old techniques are demonstrated and revived, thus honoring the tradition of craftsmanship.

But he can also take his inheritance in a completely different way. He tears down the old workshop and invests in new machines, perhaps even computer-controlled ones. He specializes in the production of office furniture and brings the business to a medium-sized level, producing his furniture efficiently and skillfully adapting to the market.

A third way would be a solution where he also renovates the business, but still keeps the old machines, continues to use them for certain purposes, and possibly discovers how they can be further developed here and there and can still be used efficiently for new materials in a new context. With the development of new building materials, the old machines suddenly find new, perhaps even surprising, uses, and he may be able to apply for a patent by modifying his machinery.

Of course, the above representation is exaggerated. However, it makes clear that historically transmitted methods do not necessarily have to be discarded to ensure the survival of a guild, be it that of carpenters or mathematicians, and that newly revived historical approaches can suddenly yield unexpectedly interesting results.

This can be, for example, a particularly simple justification for an otherwise known result. This gives the method presented here a special didactic significance.

The present project asks, on the one hand, about the potential in historical approaches that has not yet been released. This is first and foremost about awareness of the existence of alternative approaches, which cannot be assumed in young students. Secondly, it is about tackling new examples with the alternative approaches and showing that the forgotten methods often make surprisingly simple solutions to certain problems possible. Finally, it is also about comparing with standard methods, in terms of efficiency and scope. The latter aspect is of particular didactic importance. As soon as you have different methods for solving a problem in front of you, you can start comparing the methods. This is especially important in teaching. Here, differences can arise in terms of the prior knowledge of the students, the complexity of the concepts can be compared, the scope of the methods and their validity can be contrasted, and this also makes it possible to choose different paths in lesson planning. Without an awareness of the diversity of approaches, there are no choices.

Target Group

I envision the first semesters at the university as the target group. The book aims to give prospective teachers a look beyond the usual teaching content and provide content suggestions for working with interested and talented students at the secondary level.

The selection of topics is eclectic and is based on the extent to which the individual topics still contain a potential to be discovered. The topic of integration is taken up several times because there were naturally many historical approaches to the solution. The interested reader can go in search of similar historical approaches that have not yet been fully incorporated into the standardized canon of textbooks and try to further develop them themselves. Personally, I find this type of research very fruitful and important for mathematical concept formation.

In such a project, namely to take up historical methods and look for residual potential that has not yet been explored, we naturally run the risk of slipping into a certain know-it-all role. It can sometimes give the impression that we want to present methods that were developed by Archimedes or Leibniz, for example, as inadequate and want to try to improve them. This is not the intention of this book project. On the contrary, the starting point for the book project is a great admiration for the historical achievements and the intellectual work behind them. The fact that these historical results can still show new sides and be further developed only speaks for the fact that they were even stronger and more powerful than we initially assumed.

Acknowledgments

With a book like this, there are always many who have contributed and accordingly deserve a big thank you. On the one hand, good colleagues have read the individual chapters and corrected and improved them with many comments. Here I want to mention Rainer Kaenders (University of Bonn), Stefan Berendonk (University of Wuppertal), Hans Stefan Siller (University of Würzburg), Ellen Berle (Marine School Bergen, Norway) and Harald Totland (Marine School Bergen, Norway). But I also want to mention colleagues with whom I have developed the material in various articles in advance. Here my colleagues Helmer Aslaksen (University of Oslo, Norway) and Rainer Kaenders (University of Bonn) deserve special thanks. I would also like to thank Ysette Weiss (University of Mainz) for suggestions and good discussions about the book project. Without the help of the colleagues mentioned, this book project would not have been realized. I would also like to thank my wife, Grethe Nina Hestholm, for her patience with me during the writing phase.

This book is a translation of the original German edition, **"Seitenwege in der Mathematikgeschichte" (2024)**. The translation was carried out with the help of an artificial-intelligence-based machine-translation tool. My good colleague and friend Helmer Aslaksen (University of Oslo) spent weeks and months carefully reviewing the translation and provided many valuable suggestions for improvement. I owe him my deepest gratitude.

In addition, the illustrations underwent a thorough revision.

Once again, many thanks to all those mentioned here.

Bergen, Norway Christoph Kirfel

Bibliography

Aslaksen H., Kirfel C. (2024), Integration by Riemann triangles, Mathematics Magazine (Mathematical Association of America), Vol. 97, No 1, 2024, pp. 4–22.

Datta, B. (1932). *The Science of the Sulbas: A Study in Early Hindu Geometry.* Calcutta, Calcutta University Press, pp. 195–206.

Fermat P. (1896): On the transformation and simplification of locus equations, for the comparison in all forms of curvilinear areas, either among themselves, or with rectilinear ones, and at the same time on the use of geometric progression for the quadrature of parabolas and hyperbolas to infinity. In: Tannery, P., Henry C. Works of Fermat, Volume three, Gauthier-Villars et fils, 216–237. English translation: http://science.larouchepac.com/fermat/.

Grattan-Guinness, I., History or Heritage? An Important Distinction in Mathematics and for Mathematics Education, *The American Mathematical Monthly*, Vol. 111, No. 1 (Jan., 2004), pp. 1–12. https://doi.org/10.2307/4145010.

Gregorius a San Vincentio (1647). Problema austriacum plus ultra quadratura circuli. Antwerp, Mersius. (Opus geometricum quadratura circuli et sectionum coni).

Henderson, D. W. (2000). Square roots in the Sulba Sutra. In C.A. Gorini (ed): *Geometry at Work: Papers in Applied Geometry*, MAA Notes Number 53, pp. 39–45.

Kaenders, R. & Kirfel, C. (2016): Weiterentwicklung historischer Zugänge zur Infinitesimalrechnung über Elementargeometrie, Beiträge zum Mathematikunterricht 2016, hrsg. v. Institut für Mathematik und Informatik der Pädagogischen Hochschule Heidelberg. Münster WTM-Verlag, S. 1235–1238.

Kaenders, R. & Kirfel, C. (2017): Flächenbestimmung bei Basisfunktionen der Schule mit Elementargeometrie. Math. Semesterberichte, S. 199–220. 10.1007/s00591-017-0191-6.

Kaenders, R. & Kirfel, C. (2020A): Ableitung und Integral bei Basisfunktionen der Schule mit Elementargeometrie (Teil I), MNU, Jahrgang 73, 2/2020, S. 156–62.

Kaenders, R. & Kirfel, C. (2020B): Ableitung und Integral bei Basisfunktionen der Schule mit Elementargeometrie (Teil II), MNU, Jahrgang 73, 4/2020, S. 328–333.

Kirfel, C. (2013A), Arkimedes og parabelen, TANGENTEN (4/2013), Tidsskrift for matematikkundervisning, S. 46–49.

Kirfel, C. (2013B) A generalisation of Archimedes' method, *The Mathematical Gazette*, Vol. 97, No. 538 (March 2013), pp. 43–52. https://www.jstor.org/stable/24496758.

Kirfel, C. (2014A), Generalisering av Arkimedes' metode, TANGENTEN (1/2014) Tidsskrift for matrematikkundervisning. pp. 34–39.

Kirfel, C., (2014B): *Integration by geometrical means – a unified approach.* Mathematics Teaching 239. pp. 23–25.

Kirfel C. (2015). *Røtter.* Tangenten – tidsskrift for matematikkundervisning, (1/2015), pp. 36–39.

Kirfel, C. (2017A): FERMATS Methoden zur Integration, Der Mathematikunterricht, Friedrich Verlag (2017/4), pp. 44–50.

Kirfel, C. (2017B): Generalisierung der Archimedischen Methode zur Flächenbestimmung des Parabelsegments, Beiträge zum Mathematikunterricht 2017, U. Kartenkamp & Kuzle (Hrsg.) Beiträge zum Mathematikunterricht 2017, Münster WTM-Verlag, pp. 541–544.

Kirfel, C. (2018): Die logarithmische Spirale – ein dankbares Studienobjekt, Beiträge zum Mathematikunterricht 2018, Gemeinsame Jahrestagung der GDM und der DMV Paderborn, März 2018, Münster WTM-Verlag, pp. 955–958.

Kirfel, C. (2019): Indische Wurzeln – Wurzelziehen mit der Sulbasutra, Beiträge zum Mathematikunterricht 2019, A. Frank, S. Krauss & K. Binder (Hrsg.), Jahrestagung der GDM Regensburg 2019, Münster WTM-Verlag, pp. 413–416.

Kirfel, C. (2020): Die Fächermethode zur Bestimmung von Integralen, Siller, H.-S., Weigel, W. & Wörler, J. F. (Hrsg.). Beiträge zum Mathematikunterricht 2020. Münster: WTM-Verlag, 2020, S. 497–500. doi: 10.37626/GA9783959871402.0.

Kirfel, C. (2021): Indische Wurzeln, MNU (6/2021), pp. 470–478.

Kirfel, C. (2023): Die Fransenmethode zur Berechnung von Flächen, In: IDMI-Primar Goethe-Universität Frankfurt (Hrsg.) WTM, 56. Jahrestagung der Gesellschaft für Didaktik der Mathematik, Beiträge zum Mathematikunterricht 2022, pp. 917–920. https://doi.org/10.37626/GA9783959872089.0.

Kirfel, C. (2024): Die geometrische Umkehrung der Resektenmethode von Leibniz, In: Vom Mittelater in die Moderne – Episoden aus der Mathematikgeschichte, Tagungsband der Konferenz für die Geschichte der Mathematik 2023, Leipzig, WTM-Verlag, Münster, pp. 103–116.

Leibniz, G. W. (2016): De quadratura arithmetica circuli ellipseos et hyperbolae. Herausgegeben v. Eberhard Knobloch. Klassische Texte der Wissenschaft. Berlin, Heidelberg: Springer Spektrum 2016.

Competing Interests The author has no competing interests to declare that are relevant to the content of this manuscript.

Contents

List of Figures

List of Tables

Area Determinations with Archimedes

1

A Generalization of the Archimedean Method for the Quadrature of the Parabolic Segment

1.1 Introduction

In geometry, we often deal with plane figures. Triangles, quadrilaterals, circles, and curves of higher order were already of mathematical interest in antiquity. In particular, the area of these figures was of interest. For areas bounded by straight lines, such as quadrilaterals or other polygons, this can be determined using elementary methods from geometry. It becomes more difficult with areas bounded by curved lines, such as a circle. Another curved line that was known in antiquity is the parabola. Today we know this curve from "satellite dishes", the paraboloids, which adorned the roofs in the big cities until a few years ago. The shape of the parabola was already known to Archimedes. Legend has it that he constructed large parabolic mirrors and focused the sunlight on the Roman ships with them, causing them to start burning. However, many doubt the truth of this legend. Be that as it may, the shape of the parabola was a familiar object to him, and he dealt intensively with it. In particular, he was able to calculate the area of a parabolic segment that is created when a parabola is "cut off" by a straight cut. This chapter now deals with how Archimedes succeeded in this and how this method can be further generalized.

1.2 Historical Background

As early as the third century BCE, Archimedes developed a method for calculating the area of a parabolic segment. This result was unique in the early history of mathematics. His ideas and the method he used were unfortunately not further developed by his contemporaries and intellectual successors, and it took almost 1800 years until new generations of mathematicians achieved results that can be compared with Archimedes's method. In this chapter, Archimedes's method is presented in modern notation and with modern terminology. It turns out that the

© The Author(s), under exclusive license to Springer-Verlag GmbH, DE, part of Springer Nature 2026

C. Kirfel, *Side Paths in the History of Mathematics*, Mathematics Study Resources 21,
https://doi.org/10.1007/978-3-662-72918-2_1

method can be generalized to curves of higher order. Many of the arguments that Archimedes used can be directly adopted without the need for special adaptation, which shows us the strength of the Archimedean approach.

Howard Eves writes in his "Introduction to the History of Mathematics" (Eves 1983, p. 293):

> "The theory of integration received very little stimulus after Archimedes's remarkable achievements until relatively modern times. It was about 1450 that Archimedes' works reached Western Europe through a translation of a ninth-century copy of his manuscripts found at Constantinople. This translation was revised by Regiomontanus and was printed in 1540. A few years later a second translation appeared. But it was not until about the beginning of the seventeenth century that we find Archimedes's ideas receiving further development."

The inspiration for this chapter comes from the article "Integration, a genetic introduction" (Burn 1999) and Otto Toeplitz's book "The Development of Infinitesimal Calculus" (Toeplitz 1949). Parts of the material in this chapter can also be found in Kirfel (2013a, 2013b, 2014a, 2017b). For Archimedes's method, see also [Archimedes].

Archimedes developed his method to calculate the area of a parabolic segment. We will see in this chapter that the method is much more general and that it can also be used to calculate similar areas for higher curves without the need for major adjustments.

First, we want to present Archimedes's method. Our presentation does not exactly follow the path that Archimedes took. We use a modern coordinate system, the concept of function, and the notation of modern algebra to make the method easily accessible to today's readers. Above all, we want to consider power functions, i.e., $f(x) = x^n$. The parabola $f(x) = x^2$ is such a power function. The main concern of this chapter is not the exact historical reconstruction of the method, but to show the untapped potential it can have in modern thinking.

Archimedes wanted to calculate the area enclosed by a secant, a straight line connecting two points A and B on the parabola, and the parabola itself (see Fig. 1.1).

In his argument, he needs the area of the triangle ABM, where the third point M also lies on the parabola, so that it—seen in x-direction—is halfway between A and B. The area T_1 of this triangle is, so to speak, the first approximation to the area of the parabolic segment (Fig. 1.2).

Let us give the point M the coordinates $M = (m, f(m))$. Then we can write $A = (m - \Delta, f(m - \Delta))$ and $B = (m + \Delta, f(m + \Delta))$. The variable m then denotes the position of the "center", while Δ indicates half the distance between A and B in x-direction. Archimedes then first calculated the area T_1 of the triangle ABM. We will do the same here, but with modern tools.

Let H be the midpoint of the line segment AB. Then the length of the line segment MH is given by

$$MH = \frac{f(m + \Delta) + f(m - \Delta)}{2} - f(m).$$

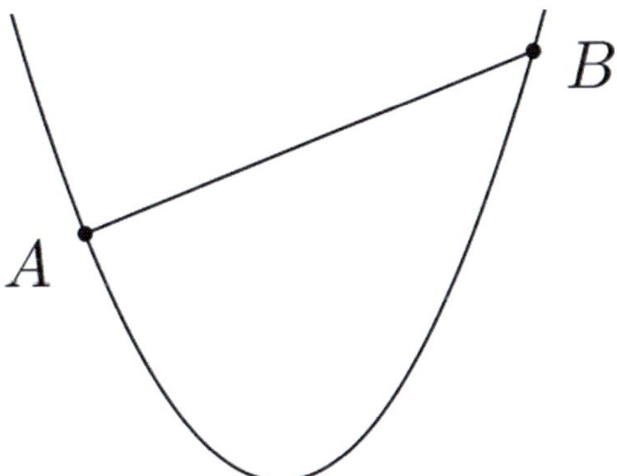

Fig. 1.1 The segment of a parabola

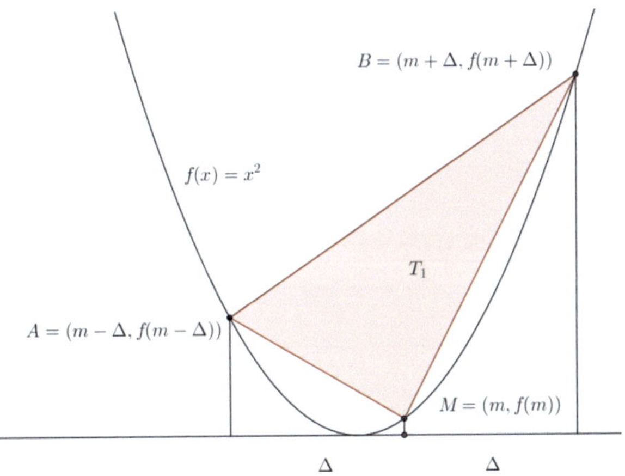

Fig. 1.2 The first triangle in the segment of a parabola

We now consider the line segment MH as the base for the triangle AMH. Then the height is Δ. The same applies to the triangle BHM and their areas are therefore equal (see Fig. 1.3). Therefore, the area T_1 of the triangle ABM is calculated as

$$T_1 = \Delta\left(\frac{f(m+\Delta) + f(m-\Delta)}{2} - f(m)\right) \tag{1.1}$$

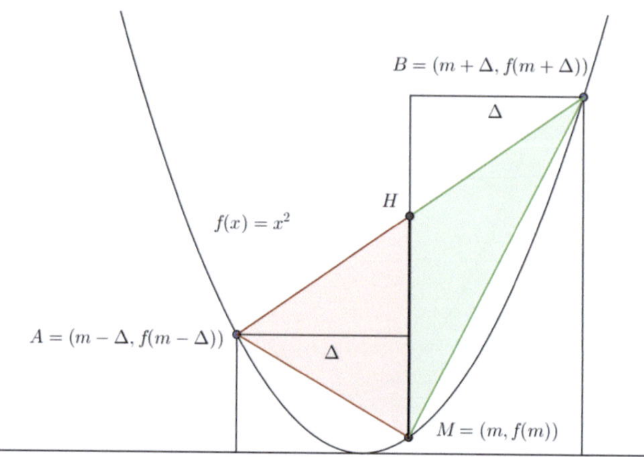

Fig. 1.3 Calculating the area of the first triangle

1.3 The Parabola

In the case of Archimedes's parabolic segment, we can write $f(x) = x^2$ and thus we obtain

$$\begin{aligned} T_1 &= \Delta\left(\frac{f(m+\Delta)+f(m-\Delta)}{2} - f(m)\right) \\ &= \Delta\left(\frac{m^2+2m\Delta+\Delta^2+m^2-2m\Delta+\Delta^2}{2} - m^2\right) \\ &= \Delta^3. \end{aligned}$$

The area T_1 of the triangle thus depends solely on half the distance Δ between A and B when we measure along the x-axis and not on where the midpoint M is located. This is quite astonishing in itself.

Archimedes's ingenious idea was now to place a new triangle T_2 between A and M. Again, the third point is chosen halfway between the other two (in x-direction). The same happens between M and B. We call this triangle T_3. But now, since the distance (in x-direction) of the outer points for the new triangle is exactly half the distance of the old one, we can easily calculate the areas (see Fig. 1.4)

$$T_2 = T_3 = \left(\frac{\Delta}{2}\right)^3 = \frac{\Delta^3}{8} \text{ and thus } T_2 + T_3 = \frac{\Delta^3}{4} = \frac{T_1}{4}.$$

Afterwards, Archimedes continues to fill in smaller and smaller triangles to fill more and more of the parabolic segment. In modern terminology, we get an infinite

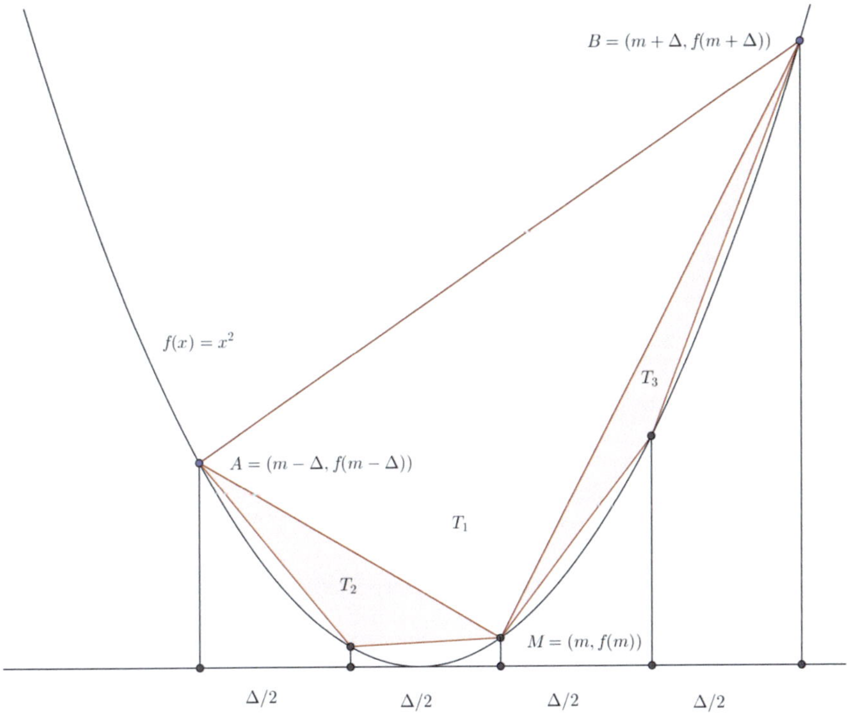

Fig. 1.4 Filling up the segment by smaller triangles

series for the total area P of the parabolic segment

$$\begin{aligned} P &= T_1 + (T_2 + T_3) + (T_4 + T_5 + T_6 + T_7) \\ &\quad + (T_8 + T_9 + T_{10} + T_{11} + T_{12} + T_{13} + T_{14} + T_{15}) + \cdots \\ &= T_1 + \left(\frac{T_1}{4}\right) + \left(\frac{T_2}{4} + \frac{T_3}{4}\right) + \left(\frac{T_4}{4} + \frac{T_5}{4} + \frac{T_6}{4} + \frac{T_7}{4}\right) + \ldots \\ &= T_1 + \frac{P}{4}. \end{aligned}$$

Here we have replaced $T_2 + T_3$ with $T_1/4$. Next, we replaced $T_4 + T_5$ with $T_2/4$ and $T_6 + T_7$ with $T_3/4$. This can be continued indefinitely. The total result is then

$$3P/4 = T_1$$

and thus

$$P = 4T_1/3 = 4\Delta^3/3.$$

Archimedes proceeded somewhat differently at this point. He constructed two contradictions. First, he assumed that $P < 4T_1/3$ and showed that this led to a contradiction. Then he assumed that $P > 4T_1/3$ and again showed that this led to a contradiction. The last possibility that then remains is that $P = 4T_1/3$, which is what we want to show.

Let's start with the first assumption, namely that $P < 4T_1/3$. The finite partial sum,

$$T_1 + \frac{T_1}{4} + \frac{T_1}{16} + \frac{T_1}{64} + \cdots + \frac{T_1}{4^n} < P,$$

because all the mentioned triangles are properly contained within the parabolic segment. However, we know the value of the finite partial sum of a geometric series—and this was also known to Archimedes. It is

$$T_1 + \frac{T_1}{4} + \frac{T_1}{16} + \frac{T_1}{64} + \cdots + \frac{T_1}{4^n} = T_1\left(\frac{1 - \left(\frac{1}{4}\right)^{n+1}}{1 - \frac{1}{4}}\right) = \frac{4}{3}T_1 - \frac{T_1}{3 \cdot 4^n} < P.$$

If the area of the entire segment were strictly smaller than $4T_1/3$, then we could find a number n such that $\frac{4}{3}T_1 - \frac{T_1}{3 \cdot 4^n} > P$, because the term $\frac{T_1}{3 \cdot 4^n}$ can be chosen arbitrarily small and thus a finite sum of triangle areas would be strictly greater than P, which of course cannot be, since all triangles are contained in the segment. Thus, the first contradiction was reached and Archimedes had to reject the assumption $P < 4T_1/3$.

Archimedes then assumed that $P > 4T_1/3$. Previously, Archimedes had shown that the tangent at the midpoint M is parallel to the secant AB (see Fig. 1.5). However, we will later show that we do not need this fact, so we will skip the proof. Thus, the quadrilateral $ABFE$ is a parallelogram. It contains the entire parabolic segment (and a bit more). This of course also applies to the further filling steps. This parallelogram, which is formed by the secant, the tangent through the midpoint and the vertical boundary lines, encompasses the entire associated parabolic segment. At the same time, the parallelogram is twice as large in terms of area as the associated triangle. If instead of adding the triangles in a step of the exhaustion process, we add the corresponding parallelograms with twice the area, we will cover more than the segment. We therefore start with a filling where initially only the triangles are covered

$$T_1 + \frac{T_1}{4} + \frac{T_1}{16} + \frac{T_1}{64} + \cdots + \frac{T_1}{4^n},$$

then we add as much area as was added in the last step, i.e. $\frac{T_1}{4^n}$. This brings the filled area to

$$T_1 + \frac{T_1}{4} + \frac{T_1}{16} + \frac{T_1}{64} + \cdots + \frac{T_1}{4^n} + \frac{T_1}{4^n} = \frac{4}{3}T_1 - \frac{T_1}{3 \cdot 4^n} + \frac{T_1}{4^n} = \frac{4}{3}T_1 + \frac{2T_1}{3 \cdot 4^n}.$$

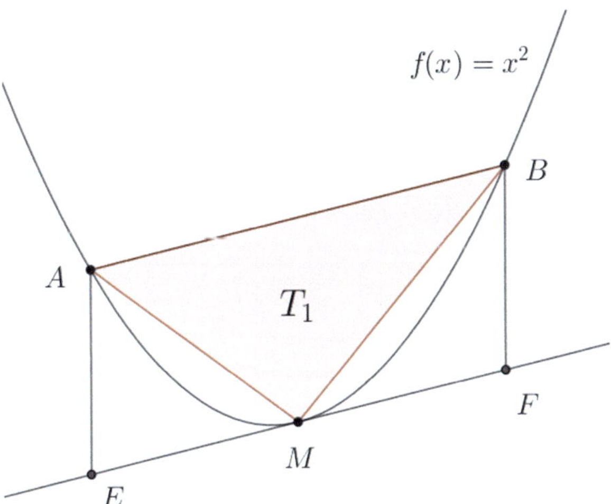

Fig. 1.5 A parallelogram containing the segment of the parabola

If the area of the parabolic segment were now $P > \frac{4}{3}T_1$, then we could find a number n so that the area with the parallelograms, i.e. $\frac{4}{3}T_1 + \frac{2T_1}{3 \cdot 4^n}$ would be smaller than P, which is impossible, since the area with the parallelograms is always larger than P. Thus, Archimedes had reached the second contradiction. So only the third alternative remained, namely that $P = \frac{4}{3}T_1$, which is what we wanted to show.

We can also easily convince ourselves in another way that the result is correct, if we consider an equilateral triangle with an area of 1 (Fig. 1.6).

On the "first floor" a quarter of the triangle is marked in dark. Two more quarters—also on the first floor—are marked in light. In the second floor it is the same.

Fig. 1.6 Division of an equilateral triangle

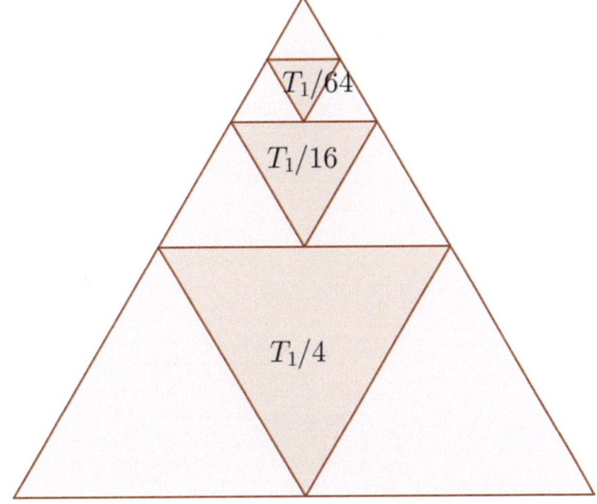

And so it goes on in each floor: always two light and one dark triangle. If we now imagine the totality of the dark triangles, all standing on the tip, they make up a third of the triangle area. So a third of the triangle is colored dark. We can therefore easily recognize that

$$\frac{1}{4} + \frac{1}{16} + \frac{1}{64} + \cdots + \frac{1}{4^n} + \cdots = \frac{1}{3}$$

and thus also

$$1 + \frac{1}{4} + \frac{1}{16} + \frac{1}{64} + \cdots + \frac{1}{4^n} + \cdots = \frac{4}{3}.$$

Therefore, we have

$$T_1 + \frac{T_1}{4} + \frac{T_1}{16} + \frac{T_1}{64} + \cdots + \frac{T_1}{4^n} + \cdots = \frac{4}{3}T_1.$$

We now return to Archimedes's result and compare it with what we know from modern integral calculus. In modern integral calculus, we are usually interested in the area between the graph of a function, the x-axis and two vertical lines, e.g. at zero and x (see Fig. 1.7). Here we have chosen $m = \Delta = \frac{x}{2}$.

Then $P = \frac{4}{3}T_1 = \frac{4}{3}\Delta^3 = \frac{4}{3}\left(\frac{x}{2}\right)^3 = \frac{x^3}{6}$, while $R = \int_0^x t^2 dt = \frac{1}{3}x^3$. Together, the two areas P and R form a triangle with the base x and the height x^2. Therefore, the total area is $\frac{1}{2}x^3$, which confirms the result of Archimedes using modern integral calculus.

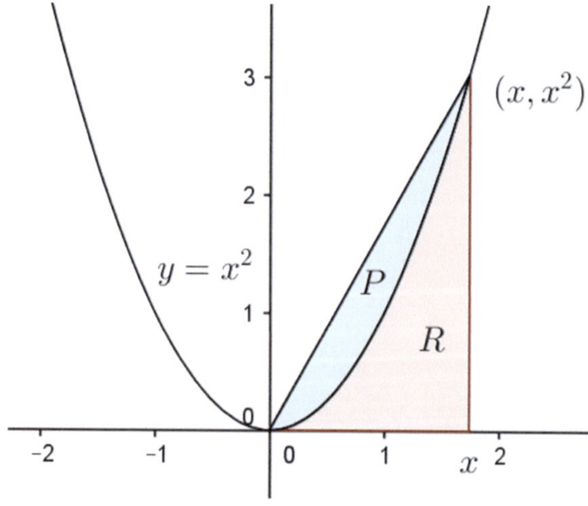

Fig. 1.7 Combining the segment of the parabola and the area under the curve

The Archimedean method in brief:

1. Choose a secant that, together with the graph of a curve, bounds a piece of area. The goal is to determine the area.
2. Place a third point on the graph halfway between the endpoints of the secant.
3. Calculate the area of the triangle.
4. Fill the curve segment with ever smaller triangles formed in the same way.
5. Determine the area as the limit of the sequence of triangle areas. Archimedes used a double contradiction proof here.

It is interesting that Archimedes's idea can be generalized to higher power functions without major adjustments. The argument with the double contradiction can also be extended to much more general functions.

To avoid negative x values, negative y values or even "negative areas"—especially with power functions with odd exponents—we will only consider functions in the first quadrant. We therefore require $0 < \Delta \leq m$ for the rest of this chapter.

1.4 The Cubic Power Function

We now examine $f(x) = x^3$. The formula (1.1) for the first approximation triangle then gives us:

$$
\begin{aligned}
T_1 &= \Delta\left(\frac{f(m+\Delta)+f(m-\Delta)}{2} - f(m)\right) \\
&= \Delta\left(\frac{m^3+3m^2\Delta+3m\Delta^2+\Delta^3+m^3-3m^2\Delta+3m\Delta^2-\Delta^3}{2} - m^3\right) \\
&= 3m\Delta^3
\end{aligned}
$$

Here, the area also depends on where A and B (and thus also M) are located and not just on half the distance, Δ, between the two. However, if we calculate the areas in the next "generation", i.e., T_2 and T_3 we get

$$
T_2 = 3\left(m - \frac{\Delta}{2}\right)\left(\frac{\Delta}{2}\right)^3 \text{ and } T_3 = 3\left(m + \frac{\Delta}{2}\right)\left(\frac{\Delta}{2}\right)^3,
$$

because the midpoint for the triangle T_2 has the x coordinate $m - \Delta/2$, while the midpoint for the triangle T_3 has the x coordinate $m+\Delta/2$. The results are different and it seems that the method is stuck here. But we are lucky. Only the sum of the two areas is of interest for the further course, and this results in

$$
T_2 + T_3 = 3\left(m - \frac{\Delta}{2}\right)\left(\frac{\Delta}{2}\right)^3 + 3\left(m + \frac{\Delta}{2}\right)\left(\frac{\Delta}{2}\right)^3 = \frac{3m\Delta^3}{4} = \frac{T_1}{4},
$$

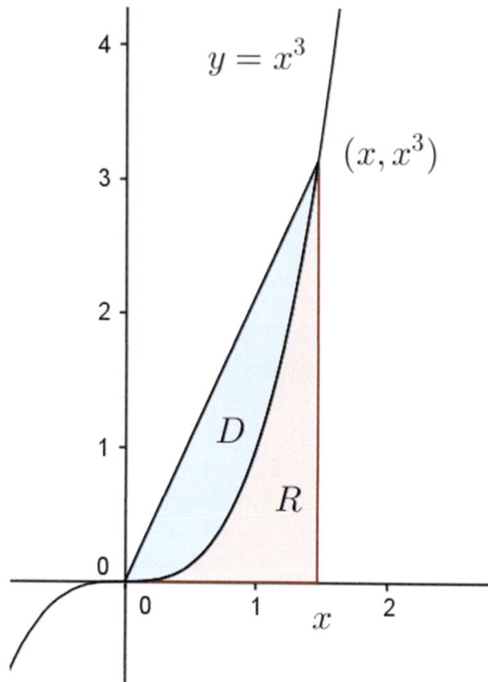

Fig. 1.8 Combining the segment of the cubic and the area under the curve

and we find ourselves in the same situation as with the parabola. The area of the triangles of the new generation is just a quarter of the area of the preceding generation. Then the final result can be calculated similarly to Archimedes' result. Let us call the area of the segment of the cubic power function D, so that

$$D = \frac{4}{3}T_1 = 4m\Delta^3.$$

Thus, we can calculate any segment of the curve $y = x^3$ as soon as m and Δ are known. Here too, we can compare the result with what modern integral calculus provides (see Fig. 1.8). Again, we set $m = \Delta = \frac{x}{2}$. Then $D = \frac{4}{3}T_1 = 4m\Delta^3 = 4\left(\frac{x}{2}\right)^4 = \frac{x^4}{4}$, while $R = \int_0^x t^3 dt = \frac{1}{4}x^4$. Together, the two areas D and R form a triangle this time with the base x and the height x^3. Therefore, the total area is $\frac{1}{2}x^4$, which confirms the result obtained with the generalized method of Archimedes.

1.5 The Power Function of Degree Four

We now consider a power function $f(x) = x^4$ of degree four and try to apply Archimedes's method to this case. Using formula (1.1) for the first approximation triangle, we obtain

$$\begin{aligned}T_1 &= \Delta\left(\frac{f(m+\Delta)+f(m-\Delta)}{2} - f(m)\right)\\&= \Delta\left(\frac{m^4+4m^3\Delta+6m^2\Delta^2+4m\Delta^3+\Delta^4+m^4-4m^3\Delta+6m^2\Delta^2-4m\Delta^3+\Delta^4}{2} - m^4\right)\\&= 6m^2\Delta^3+\Delta^5 = \Delta^3\left(6m^2+\Delta^2\right).\end{aligned}$$

Here, the area again depends on both m, that is, where the intermediate point M is located and half the distance Δ between the endpoints of the secant. If we calculate the area of the next generation T_2 and T_3, we obtain

$$T_2 = \left(\frac{\Delta}{2}\right)^3\left(6\left(m-\frac{\Delta}{2}\right)^2+\left(\frac{\Delta}{2}\right)^2\right) \quad \text{and} \quad T_3 = \left(\frac{\Delta}{2}\right)^3\left(6\left(m+\frac{\Delta}{2}\right)^2+\left(\frac{\Delta}{2}\right)^2\right)$$

The sum this time yields

$$\begin{aligned}T_2+T_3 &= \left(\frac{\Delta}{2}\right)^3\left(6\left(m-\frac{\Delta}{2}\right)^2+\left(\frac{\Delta}{2}\right)^2+6\left(m+\frac{\Delta}{2}\right)^2+\left(\frac{\Delta}{2}\right)^2\right)\\&= \left(\frac{\Delta}{2}\right)^3\left(12m^2+3\Delta^2+\frac{\Delta^2}{2}\right)\\&= \frac{\Delta^3}{4}\left(6m^2+\frac{7\Delta^2}{4}\right)\\&= \frac{\Delta^3}{4}(6m^2+\Delta^2)+\frac{3\Delta^5}{16} = \frac{T_1}{4}+\frac{3\Delta^5}{16}.\end{aligned}$$

In addition to $T_1/4$ we get an extra term $3\Delta^5/16$, which did not appear for the parabola and the function of degree three and initially gives us some headache.

However, we soon see that both the term $\frac{T_1}{4}$ and the term $\frac{3\Delta^5}{16}$ each lead to geometric series that we already know, and that Archimedes was also acquainted with.

Now we use the letter V for the segment of the fourth degree power function. Then we can write

$$\begin{aligned}V &= T_1+(T_2+T_3)+(T_4+T_5+T_6+T_7)+\\&\quad(T_8+T_9+T_{10}+T_{11}+T_{12}+T_{13}+T_{14}+T_{15})+\cdots\\&= T_1+\left(\frac{T_1}{4}+\frac{3\Delta^5}{16}\right)+\left(\frac{T_2}{4}+\frac{3\Delta^5}{16}\left(\frac{1}{2}\right)^5+\frac{T_3}{4}+\frac{3\Delta^5}{16}\left(\frac{1}{2}\right)^5\right)\\&\quad+\left(\frac{T_4}{4}+\frac{3\Delta^5}{16}\left(\frac{1}{4}\right)^5+\frac{T_5}{4}+\frac{3\Delta^5}{16}\left(\frac{1}{4}\right)^5\right.\end{aligned}$$

$$
\begin{aligned}
&+\frac{T_6}{4}+\frac{3\Delta^5}{16}\left(\frac{1}{4}\right)^5+\frac{T_7}{4}+\frac{3\Delta^5}{16}\left(\frac{1}{4}\right)^5\Bigg)+\cdots \\
&=T_1+\frac{V}{4}+\frac{3\Delta^5}{16}\left(1+2\left(\frac{1}{2}\right)^5+4\left(\frac{1}{4}\right)^5+\cdots\right) \\
&=T_1+\frac{V}{4}+\frac{3\Delta^5}{16}Z.
\end{aligned}
$$

Here, Z refers to the expression inside the last big parenthesis, which is a geometric series

$$
\begin{aligned}
Z&=1+2\left(\frac{1}{2}\right)^5+4\left(\frac{1}{4}\right)^5+\cdots=1+\left(\frac{1}{2}\right)^4+\left(\frac{1}{4}\right)^4+\cdots \\
&=1+\frac{1}{16}+\frac{1}{16^2}+\cdots=\frac{16}{15}.
\end{aligned}
$$

This is not how Archimedes would have argued. However, he also knew geometric series and could compute them.

Substituting this value for Z into the expression for V, we get that the area of the segment is

$$
V=T_1+\frac{V}{4}+\frac{\Delta^5}{5},
$$

which brings us the following

$$
V=\frac{4}{3}\left(T_1+\frac{\Delta^5}{5}\right)=\frac{4}{3}\left(\Delta^3(6m^2+\Delta^2)+\frac{\Delta^5}{5}\right)=\frac{8\Delta^3(5m^2+\Delta^2)}{5}.
$$

If we again set $m=\Delta=\frac{x}{2}$, we then get

$$
V=\frac{8\Delta^3(5m^2+\Delta^2)}{5}=\frac{3x^5}{10}.
$$

When compared with the result from integral calculus, we get $R=\int_0^x x^4dt=\frac{x^5}{5}$. Together with the curve segment V we then get $V+R=\frac{3x^5}{10}+\frac{x^5}{5}=\frac{5x^5}{10}=\frac{x^5}{2}$. Again, the two area pieces fill the triangle with base x and this time height x^4 (see Fig. 1.9). Thus, modern integral calculus has once again been able to confirm the results of the extended Archimedean method.

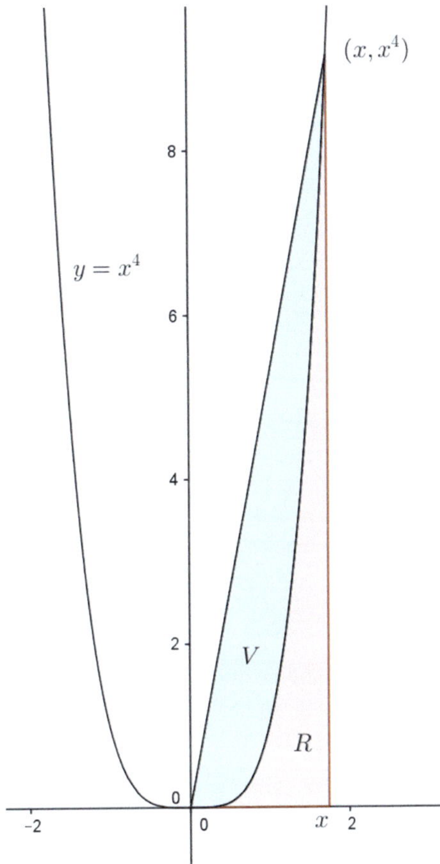

Fig. 1.9 Combing the segment of the quartic and the area below the curve

1.6 The Power Function of Degree Five

We now consider a power function $f(x) = x^5$ of degree five. Using formula (1.1) for the first approximation triangle, we obtain this time

$$\begin{aligned}
T_1 &= \Delta\left(\frac{f(m+\Delta) + f(m-\Delta)}{2} - f(m)\right) \\
&= \Delta\left(\frac{(m+\Delta)^5 + (m-\Delta)^5}{2} - m^5\right) \\
&= \Delta\left(\frac{m^5 + 5m^4\Delta + 10m^3\Delta^2 + 10m^2\Delta^3 + 5m\Delta^4 + \Delta^5}{2}\right. \\
&\quad \left. + \frac{m^5 - 5m^4\Delta + 10m^3\Delta^2 - 10m^2\Delta^3 + 5m\Delta^4 - \Delta^5}{2} - m^5\right)
\end{aligned}$$

$$= 10m^3\Delta^3 + 5m\Delta^5 = 5m\Delta^3(2m^2 + \Delta^2).$$

With this, we can calculate the area of the triangles in the next generation.

$$\begin{aligned}
T_2 + T_3 &= 5\left(m - \frac{\Delta}{2}\right)\left(\frac{\Delta}{2}\right)^3\left(2\left(m - \frac{\Delta}{2}\right)^2 + \left(\frac{\Delta}{2}\right)^2\right) \\
&\quad + 5\left(m + \frac{\Delta}{2}\right)\left(\frac{\Delta}{2}\right)^3\left(2\left(m + \frac{\Delta}{2}\right)^2 + \left(\frac{\Delta}{2}\right)^2\right) \\
&= 5\left(m - \frac{\Delta}{2}\right)\left(\frac{\Delta}{2}\right)^3\left(2m^2 - 2m\Delta + \frac{3\Delta^2}{4}\right) \\
&\quad + 5\left(m + \frac{\Delta}{2}\right)\left(\frac{\Delta}{2}\right)^3\left(2m^2 + 2m\Delta + \frac{3\Delta^2}{4}\right) \\
&= \frac{5m\Delta^3}{4}\left(2m^2 + \frac{7\Delta^2}{4}\right) = \frac{5m\Delta^3}{4}\left(2m^2 + \Delta^2 + \frac{3\Delta^2}{4}\right) \\
&= \frac{T_1}{4} + \frac{15m\Delta^5}{16}.
\end{aligned}$$

We see that the position m of the intermediate point again enters into the formula for the area. If we look one generation further, it looks like this:

$$\begin{aligned}
T_4 + T_5 + T_6 + T_7 &= \frac{T_2}{4} + \frac{15}{16}\left(m - \frac{\Delta}{2}\right)\left(\frac{\Delta}{2}\right)^5 + \frac{T_3}{4} + \frac{15}{16}\left(m + \frac{\Delta}{2}\right)\left(\frac{\Delta}{2}\right)^5 \\
&= \frac{T_2 + T_3}{4} + \frac{15m\Delta^5}{16} \cdot \frac{2}{32} = \frac{T_2 + T_3}{4} + \frac{15m\Delta^5}{16} \cdot \frac{1}{16}.
\end{aligned}$$

Two of the terms cancel each other. One generation further, it looks like this:

$$\begin{aligned}
&T_8 + T_9 + T_{10} + T_{11} + T_{12} + T_{13} + T_{14} + T_{15} \\
&= \frac{T_4}{4} + \frac{15}{16}\left(m - \frac{3\Delta}{4}\right)\left(\frac{\Delta}{4}\right)^5 + \frac{T_5}{4} + \frac{15}{16}\left(m - \frac{\Delta}{4}\right)\left(\frac{\Delta}{4}\right)^5 \\
&\quad + \frac{T_6}{4} + \frac{15}{16}\left(m + \frac{\Delta}{4}\right)\left(\frac{\Delta}{4}\right)^5 + \frac{T_7}{4} + \frac{15}{16}\left(m + \frac{3\Delta}{4}\right)\left(\frac{\Delta}{4}\right)^5 \\
&= \frac{T_4 + T_5 + T_6 + T_7}{4} + \frac{15m\Delta^5}{16} \cdot \frac{4}{4^5} \\
&= \frac{T_4 + T_5 + T_6 + T_7}{4} + \frac{15m\Delta^5}{16} \cdot \frac{1}{16^2}.
\end{aligned}$$

Here, there are already four terms that together equal zero. The additions and subtractions thus constantly balance each other out and in addition to the series $\frac{T_1+T_2+T_3+T_4+\cdots}{4}$ we get the series

$$\frac{15m\Delta^5}{16}\left(1+\frac{1}{16}+\frac{1}{16^2}+\cdots+\frac{1}{16^n}+\cdots\right)=\frac{15m\Delta^5}{16}\cdot\frac{16}{15}=m\Delta^5$$

Now we use the letter F for the segment of the fifth degree power function, so we can write

$$\begin{aligned}F &= T_1+(T_2+T_3)+(T_4+T_5+T_6+T_7)\\ &\quad+(T_8+T_9+T_{10}+T_{11}+T_{12}+T_{13}+T_{14}+T_{15})+\cdots\\ &= T_1+\left(\frac{T_1}{4}+\frac{15m}{16}\Delta^5\right)+\left(\frac{T_2}{4}+\frac{15m}{16}\left(\frac{\Delta}{2}\right)^5+\frac{T_3}{4}+\frac{15m}{16}\left(\frac{\Delta}{2}\right)^5\right)\\ &\quad+\left(\frac{T_4}{4}+\frac{15m}{16}\left(\frac{\Delta}{4}\right)^5+\frac{T_5}{4}+\frac{15m}{16}\left(\frac{\Delta}{4}\right)^5\right.\\ &\quad\left.+\frac{T_6}{4}+\frac{15m}{16}\left(\frac{\Delta}{4}\right)^5+\frac{T_7}{4}+\frac{15m}{16}\left(\frac{\Delta}{4}\right)^5\right)+\cdots\\ &= T_1+\frac{F}{4}+\frac{15m}{16}\Delta^5\left(1+2\left(\frac{\Delta}{2}\right)^5+4\left(\frac{\Delta}{4}\right)^5+\cdots\right)\\ &= T_1+\frac{F}{4}+m\Delta^5\end{aligned}$$

This means

$$F=\frac{4}{3}\left(T_1+m\Delta^5\right)=\frac{4}{3}\left(5m\Delta^3\left(2m^2+\Delta^2\right)+m\Delta^5\right)=\frac{8}{3}m\Delta^3\left(5m^2+3\Delta^2\right).$$

Again, we can compare with the result from integral calculus by setting $m=\Delta=\frac{x}{2}$. Then $F=\frac{8m\Delta^3(5m^2+3\Delta^2)}{3}=\frac{8}{3}\left(\frac{x}{2}\right)^4\left(5\left(\frac{x}{2}\right)^2+3\left(\frac{x}{2}\right)^2\right)=\frac{x^6}{3}$, while

$$R=\int_0^x t^5\,dt=\frac{x^6}{6}.$$

Together, the two areas add up to $F+R=\frac{x^6}{3}+\frac{x^6}{6}=\frac{x^6}{2}$. Thus, they once again fill the triangle that has the base x and this time the height x^5 (see Fig. 1.10).

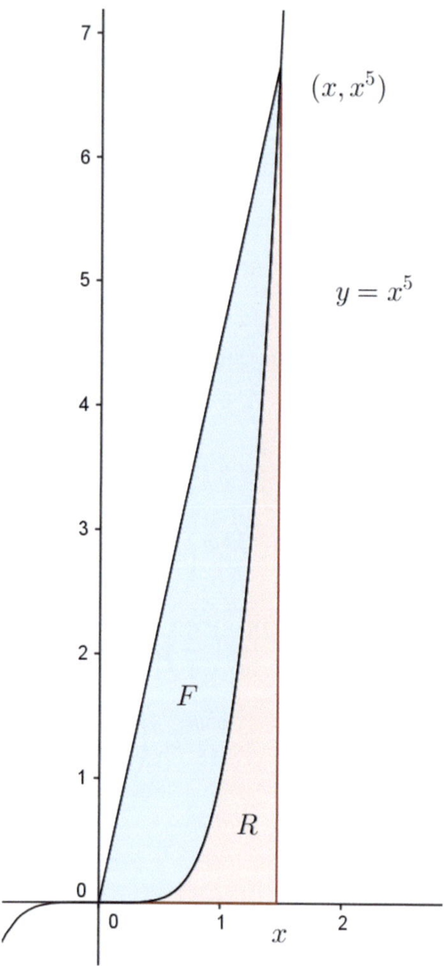

Fig. 1.10 Combining the segment of the quintic and the area below the curve

1.7 Archimedes's Argument with Double Contradiction

We now want to show how Archimedes's original argument looks for parabolas. Then we will apply it to polynomials of degree below six and finally show how we can extend it to any convex, monotonic, and differentiable curve.

We must require differentiability because we need the tangents at the intermediate points.

Archimedes used a method here that is already in the tenth book of Euclid's Elements in Proposition 1 (see [Euclid 1] and [Euclid 2]).

For the parabola (see Fig. 1.11), the triangle T_1 exactly fills half of the parallelogram $AEFB$ and thus more than half of the segment. This also applies to the next step of the exhaustion process and of course to all further ones. At each step, more than half of the remaining area is added by the new triangles. Proposition 1 from

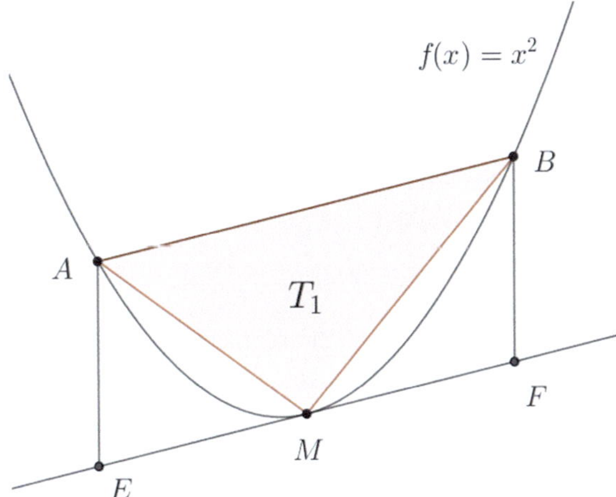

Fig. 1.11 The tangent to the parabola at the midpoint M is parallel to the secant AB

the tenth book of Euclid's Elements states that if you exhaust more than half of the remaining quantity at each step, then the remainder eventually becomes smaller than any given amount. Here it was important that the tangent at M is parallel to the secant AB, which Archimedes had shown beforehand. Thus, the quadrilateral $AEFB$ is a parallelogram and the triangle AMB is half the area.

For the parabola, Archimedes used these parallelograms for the other contradiction. These were twice as large as the corresponding triangles.

We have already seen that the sequence of triangles that Archimedes constructed up to the n-th step are all contained within the parabolic segment. If we had used parallelograms instead of triangles in the last step, which are all twice as large as the corresponding triangles, we would have obtained an area that completely encompasses the segment, thus being larger than the segment. If we again write S for the area of the segment, we have already seen that

$$
\begin{aligned}
&T_1 < S < 2T_1 \\
&T_1 + (T_2 + T_3) < S < T_1 + 2(T_2 + T_3) \\
&T_1 + (T_2 + T_3) + (T_4 + T_5 + T_6 + T_7) < S < T_1 + (T_2 + T_3) + 2(T_4 + T_5 + T_6 + T_7)
\end{aligned}
$$

Here we simplify our notation and write

$$
\begin{aligned}
G_0 &= T_1 \\
G_1 &= T_2 + T_3 \\
G_2 &= T_4 + T_5 + T_6 + T_7
\end{aligned}
$$

and in general

$$G_n = T_{2^n} + T_{2^n+1} + T_{2^n+2} + \cdots + T_{2^{n+1}-1}$$

for the total area of the triangles added in the n-th generation. Then we have

$$\begin{aligned}
&G_0 < S < 2G_0 \\
&G_0 + G_1 < S < G_0 + 2G_1 \\
&G_0 + G_1 + G_2 < S < G_0 + G_1 + 2G_2 \\
&\ldots \\
&G_0 + G_1 + G_2 + \cdots + G_n < S < G_0 + G_1 + G_2 + \cdots + 2G_n
\end{aligned}$$

etc. Thus, the area S of the segment is enclosed between two quantities, the difference of which is only G_n. We now want to try to show that G_n forms a null sequence.

In the section on the parabola, however, we saw that

$$\begin{aligned}
G_0 &= T_1 \\
G_1 &= T_2 + T_3 = \frac{T_1}{4} \\
G_2 &= T_4 + T_5 + T_6 + T_7 = \frac{T_1}{16}
\end{aligned}$$

and in general

$$G_n = T_{2^n} + T_{2^n+1} + T_{2^n+2} + \ldots + T_{2^{n+1}-1} = \frac{T_1}{4^n}.$$

Thus, the difference between the upper and lower bounds for the area S of the segment forms a null sequence, and the lower and upper bounds converge to S.

The same is true for the cubic function, since here we also have $G_n = \frac{T_1}{4^n}$.

In the case of the fourth-degree power function, it is a little more complicated but not particularly difficult. There we were able to show that

$$\begin{aligned}
G_0 &= T_1 \\
G_1 &= T_2 + T_3 = \frac{T_1}{4} + \frac{3\Delta^5}{16} \\
G_2 &= T_4 + T_5 + T_6 + T_7 = \frac{T_1}{16} + \frac{3\Delta^5}{16^2}
\end{aligned}$$

and in general

$$G_n = T_{2^n} + T_{2^n+1} + T_{2^n+2} + \cdots + T_{2^{n+1}-1} = \frac{T_1}{4^n} + \frac{3\Delta^5}{16^n}.$$

Again, the difference between the upper and lower bounds for the area S of the segment forms a null sequence, and the lower and upper bounds converge to S.

The same is true for the fifth-degree function. Here we have

$$G_0 = T_1$$
$$G_1 = T_2 + T_3 = \frac{T_1}{4} + \frac{15m\Delta^5}{16}$$
$$G_2 = T_4 + T_5 + T_6 + T_7 = \frac{T_1}{16} + \frac{15m\Delta^5}{16^2}$$

and in general

$$G_n = T_{2^n} + T_{2^n+1} + T_{2^n+2} + \cdots + T_{2^{n+1}-1} = \frac{T_1}{4^n} + \frac{15m\Delta^5}{16^n}.$$

Again, the difference between the upper and lower bounds for the area S of the segment forms a null sequence, and the lower and upper bounds converge to S.

In the case of a general convex, monotonic differentiable curve, the tangent at the intermediate point M does not need to be parallel to the secant AB (see Fig. 1.12) and we cannot assume the quadrilateral $AEFB$ to be a parallelogram. However, we see that due to the convexity, the quadrilateral $AEFB$ contains the entire segment (and a little more). EF here is the tangent at the point M on the parabola, whose x-coordinate is halfway between A and B. We have chosen points E and F so that the lines AE and BF are vertical.

Now we rotate the line EF around the intermediate point M, until it runs parallel to the secant AB. We call the intersection points with the vertical lines G and H. The new quadrilateral $AGHB$ has the same area as the previous one $AEFB$, because the lost triangle MHF is congruent to the gained triangle EMG. It was important here that the rotation point M was the midpoint of the line segment EF. Now we can see immediately that T_1 fills half of the parallelogram $AGHB$ and thus also half of $AEFB$ and thus again more than half of the segment. At each step, however, more than half of the remaining area is filled, and thus the total area is eventually filled.

We now want to show that we also obtain a null sequence in the general case of convex, monotonic, differentiable functions, not just with power functions. To do this, we look at Fig. 1.13. We see that the triangles belonging to the same generation can each be packed into rectangles of the same width $\frac{\Delta}{2^n}$. These can then be stacked on top of each other, and due to the monotonicity of the function, the total stack height is equal to the difference $y_B - y_A$ of the y-values of the endpoints of the secant on the function graph. This means that the total area of the triangles belonging to one generation is bounded by the area of the stacked rectangles, the total size of which is $\frac{(y_B - y_A)\Delta}{2^n}$ and thus represents a null sequence.

Therefore, the Archimedean method can be used for any function that is convex, monotonic, and differentiable over a finite interval. Here, the monotonicity was important so that the rectangles could be stacked into a rectangle of finite height during the stacking process. The differentiability of the function was important so that we could use the tangents at the intermediate points, and the convexity

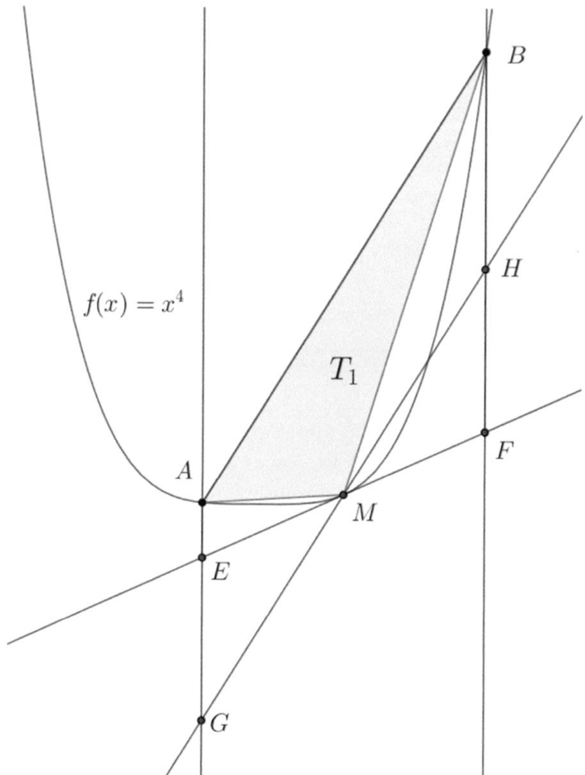

Fig. 1.12 Rotation of the tangent about the point of tangency

of the curve was important so that the tangents always lie on the same side of the curve and we could argue with the enclosed triangles and the surrounding parallelograms.

Without much effort, the result can also be extended to functions that are piecewise monotonic, convex, and differentiable, by breaking down the function into pieces that are monotonic, convex and differentiable and then applying the above result to each subinterval. Thus, all functions that students encounter during their school years meet the above conditions, and it can be said that the Archimedean method covers the functions from school mathematics when it comes to the question of integrability.

1.8 Summary

Of course, Archimedes did not have modern notation at hand, nor could he rely on the concept of a function or a coordinate system. Functions and expressions of the fourth degree or higher were hard to imagine for people from the time of

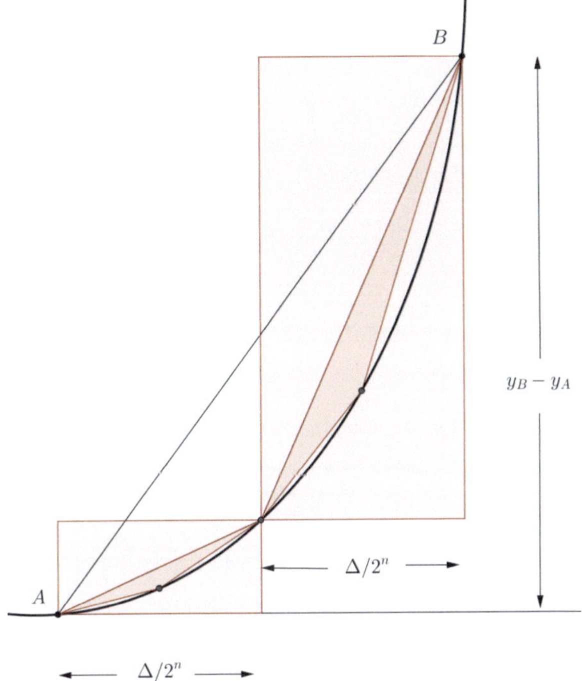

Fig. 1.13 Piling up the rectangles that contain the triangles

Archimedes, as they interpreted such expressions as lengths, areas, or volumes. Nevertheless, Archimedes's method contains in essence an idea that is not only limited to parabolas, and which can easily be applied to functions of higher order and more general form today. Also, his argument of double contradiction can readily be extended to convex, monotonic, differentiable functions and therefore not only includes the power functions discussed in detail here. Thus, the idea behind the Archimedean method already anticipates the essential aspects of the Riemann integral.

In this sense, it is justified to say that Archimedes deserves an even greater share of the credit for the development of modern integral calculus than we have so far given him.

1.9 Afterword: The Power Function of Degree Six

At this point, the reader may wonder: What about the sixth or higher powers? The sixth power unfortunately looks difficult. Here we need new tools that were not necessary before. We want to show that the Archimedean method could also be used here, but that the computations become complicated for $f(x) = x^6$.

With the help of formula (1.1) we now determine the first approximation triangle like we did for the previous power functions, and this time we obtain

$$\begin{aligned}
T_1 =& \Delta\left(\frac{f(m+\Delta)+f(m-\Delta)}{2}-f(m)\right)\\
=& \Delta\left(\frac{m^6+6m^5\Delta+15m^4\Delta^2+20m^3\Delta^3+15m^2\Delta^4+6m\Delta^5+\Delta^6}{2}\right)\\
&+\Delta\left(\frac{m^6-6m^5\Delta+15m^4\Delta^2-20m^3\Delta^3+15m^2\Delta^4-6m\Delta^5+\Delta^6}{2}\right)-\Delta m^6\\
=& 15m^4\Delta^3+15m^2\Delta^5+\Delta^7=\Delta^3\left(15m^4+15m^2\Delta^2+\Delta^4\right).
\end{aligned}$$

We can then again calculate the area in the next generation.

$$\begin{aligned}
&T_2+T_3\\
&=\left(\frac{\Delta}{2}\right)^3\left(15\left(m-\frac{\Delta}{2}\right)^4+15\left(m-\frac{\Delta}{2}\right)^2\left(\frac{\Delta}{2}\right)^2+\left(\frac{\Delta}{2}\right)^4\right)\\
&\quad+\left(\frac{\Delta}{2}\right)^3\left(15\left(m+\frac{\Delta}{2}\right)^4+15\left(m+\frac{\Delta}{2}\right)^2\left(\frac{\Delta}{2}\right)^2+\left(\frac{\Delta}{2}\right)^4\right)\\
&=\left(\frac{\Delta}{2}\right)^3\left(15\left(m^4-4m^3\frac{\Delta}{2}+6m^2\left(\frac{\Delta}{2}\right)^2-4m\left(\frac{\Delta}{2}\right)^3+\left(\frac{\Delta}{2}\right)^4+m^4\right.\right.\\
&\quad+4m^3\frac{\Delta}{2}+6m^2\left(\frac{\Delta}{2}\right)^2+4m\left(\frac{\Delta}{2}\right)^3+\left(\frac{\Delta}{2}\right)^4+m^2\left(\frac{\Delta}{2}\right)^2-2m\left(\frac{\Delta}{2}\right)^3\\
&\quad\left.\left.+\left(\frac{\Delta}{2}\right)^4+m^2\left(\frac{\Delta}{2}\right)^2+2m\left(\frac{\Delta}{2}\right)^3+\left(\frac{\Delta}{2}\right)^4\right)+2\left(\frac{\Delta}{2}\right)^4\right)\\
&=2\left(\frac{\Delta}{2}\right)^3\left(15\left(m^4+6m^2\left(\frac{\Delta}{2}\right)^2+\left(\frac{\Delta}{2}\right)^4+m^2\left(\frac{\Delta}{2}\right)^2+\left(\frac{\Delta}{2}\right)^4\right)+\left(\frac{\Delta}{2}\right)^4\right)\\
&=\frac{\Delta^3}{4}\left(15m^4+105m^2\left(\frac{\Delta}{2}\right)^2+31\left(\frac{\Delta}{2}\right)^4\right)=\frac{\Delta^3}{4}\left(15m^4+\frac{105}{4}m^2\Delta^2+\frac{31}{16}\Delta^4\right)\\
&=\frac{\Delta^3}{4}\left(15m^4+15m^2\Delta^2+\Delta^4\right)+\frac{\Delta^3}{4}\left(\frac{45}{4}m^2\Delta^2+\frac{15}{16}\Delta^4\right)\\
&=\frac{T_1}{4}+\frac{15\Delta^7}{64}+\frac{45\Delta^5m^2}{16}.
\end{aligned}$$

The first two terms again lead to known series. We have

$$\frac{T_1+T_2+T_3+\cdots}{4}=\frac{S}{4}$$

and

$$\frac{15\Delta^7}{64}\left(1+\frac{2}{2^7}+\frac{4}{4^7}+\cdots\right)=\frac{15\Delta^7}{64}\cdot\frac{64}{63}=\frac{5\Delta^7}{21}.$$

The third term $45\Delta^5 m^2/16$ is problematic, as the interval midpoints m change from step to step. This was also the case with the fifth power, but since m entered the equation linearly, the positive and negative contributions canceled each other. This is no longer the case here, because m enters the equation quadratically. So, we have to calculate the following expression:

$$\begin{aligned}K=\Bigg(&\Delta^5 m^2+\left(\frac{\Delta}{2}\right)^5\left(m-\frac{\Delta}{2}\right)^2+\left(\frac{\Delta}{2}\right)^5\left(m+\frac{\Delta}{2}\right)^2\\&+\left(\frac{\Delta}{4}\right)^5\left(m-\frac{3\Delta}{4}\right)^2+\left(\frac{\Delta}{4}\right)^5\left(m-\frac{\Delta}{4}\right)^2\\&+\left(\frac{\Delta}{4}\right)^5\left(m+\frac{\Delta}{4}\right)^2+\left(\frac{\Delta}{4}\right)^5\left(m+\frac{3\Delta}{4}\right)^2+\cdots\Bigg).\end{aligned}$$

We see that the mixed terms in the quadratic expressions cancel out. Therefore, we get

$$\begin{aligned}\mathrm{K}=&\frac{45\Delta^5 m^2}{16}\left(1+2\left(\frac{1}{2}\right)^5+4\left(\frac{1}{4}\right)^5+\cdots\right)\\&+\frac{45\Delta^7}{16}\left((1^2+1^2)\left(\frac{1}{2}\right)^7+(3^2+1^2+1^2+3^2)\left(\frac{1}{4}\right)^7+\cdots\right).\end{aligned}$$

The first expression corresponds again to a geometric series that we already know

$$\frac{45\Delta^5 m^2}{16}\left(1+2\left(\frac{1}{2}\right)^5+4\left(\frac{1}{4}\right)^5+\cdots\right)=\frac{45\Delta^5 m^2}{16}\frac{16}{15}=3\Delta^5 m^2.$$

In the second expression, twice the sum of the odd square numbers up to 2^n-1 appears,

$$\begin{aligned}&1^2\\&1^2+3^2\\&1^2+3^2+5^2+7^2.\end{aligned}$$

Now for the sum of the square numbers up to 2^n the following applies:

$$1^2 + 2^2 + 3^2 + 4^2 + 5^2 + \cdots + (2^n)^2 = \sum_{i=1}^{2^n} i^2 = \frac{2^n(2^n + 1)(2^{n+1} + 1)}{6}.$$

Here, the even square numbers make up the following part of the whole sum

$$\sum_{i=1}^{2^{n-1}} (2i)^2 = 4\sum_{i=1}^{2^{n-1}} i^2 = 4 \cdot \frac{2^{n-1}(2^{n-1} + 1)(2^n + 1)}{6}.$$

If we now subtract the even square numbers to obtain the sum of the odd ones, we get

$$\begin{aligned} 1^2 + 3^2 + 5^2 + \cdots (2^n - 1)^2 &= \frac{2^n(2^n + 1)(2^{n+1} + 1)}{6} - \frac{2^{n+1}(2^{n-1} + 1)(2^n + 1)}{6} \\ &= 2^n(2^n + 1)\left(\frac{2^{n+1} + 1}{6} - \frac{2^n + 2}{6}\right) \\ &= \frac{2^n(2^n + 1)(2^n - 1)}{6} = \frac{2^n(2^{2n} - 1)}{6}. \end{aligned}$$

For the expression

$$\frac{45\Delta^7}{16}\left((1^2 + 1^2)\left(\frac{1}{2}\right)^7 + (3^2 + 1^2 + 1^2 + 3^2)\left(\frac{1}{4}\right)^7 + \cdots\right),$$

we then obtain

$$\begin{aligned} 2\frac{45\Delta^7}{16}\sum_{n=1}^{\infty}\frac{2^n(2^{2n} - 1)}{6 \cdot 2^{7n}} &= \frac{15\Delta^7}{16}\sum_{n=1}^{\infty}\frac{2^{2n} - 1}{2^{6n}} = \frac{15\Delta^7}{16}\left(\sum_{n=1}^{\infty}\frac{1}{2^{4n}} - \sum_{n=1}^{\infty}\frac{1}{2^{6n}}\right) \\ &= \frac{15\Delta^7}{16}\left(\frac{1}{15} - \frac{1}{63}\right) = \frac{\Delta^7}{21}. \end{aligned}$$

For the total area, we can then write

$$\begin{aligned} S &= T_1 + (T_2 + T_3) + (T_4 + T_5 + T_6 + T_7) \\ &\quad + (T_8 + T_9 + T_{10} + T_{11} + T_{12} + T_{13} + T_{14} + T_{15}) \\ &= T_1 + \frac{S}{4} + \frac{5\Delta^7}{21} + 3\Delta^5 m^2 + \frac{\Delta^7}{21} = T_1 + \frac{S}{4} + 3\Delta^5 m^2 + \frac{2\Delta^7}{7}, \end{aligned}$$

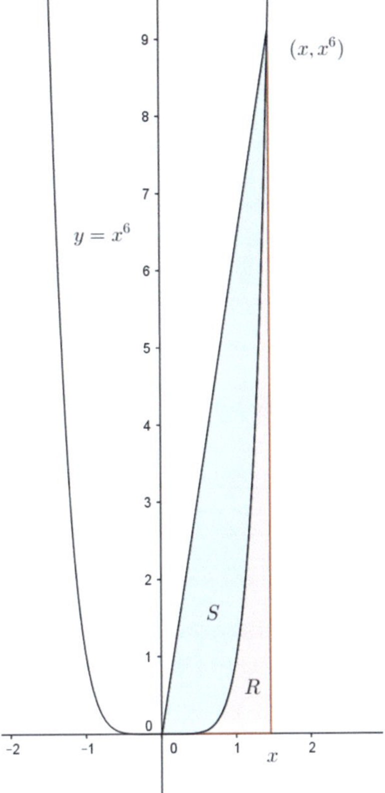

Fig. 1.14 Combining the segment of the function of degree six and the area under the curve

thus

$$S = \frac{4}{3}\left(T_1 + 3m^2\Delta^5 + \frac{2\Delta^7}{7}\right) = \frac{4}{3}\left(15m^4\Delta^3 + 15m^2\Delta^5 + \Delta^7 + 3m^2\Delta^5 + \frac{2\Delta^7}{7}\right)$$
$$= \frac{4}{7}\left(35m^4\Delta^3 + 42m^2\Delta^5 + 3\Delta^7\right).$$

Again, we can compare with the result from integral calculus by setting $m = \Delta = \frac{x}{2}$. Then $S = \frac{4}{7}\left(35m^4\Delta^3 + 42m^2\Delta^5 + 3\Delta^7\right) = \frac{4}{7}\left(\frac{x}{2}\right)^7(35 + 42 + 3) = \frac{80x^7}{7 \cdot 32} = \frac{5x^7}{14}$, while

$$R = \int_0^x t^6 dt = \frac{x^7}{7}.$$

Together, the two areas yield $S + R = \frac{5x^7}{14} + \frac{x^7}{7} = \frac{x^7}{2}$. Thus, they again jointly fill the triangle, which has baseline x and this time height x^6 (see Fig. 1.14).

At this stage, if not earlier, we see that the argument is becoming increasingly complex. In the last example, the sum of the odd square numbers appeared. Later, the sums of the odd cube numbers or even higher powers of the natural numbers would also appear. Thus, the ingredients for the proof are much harder to establish than obtaining the result using other integration methods. Therefore, it makes little sense to develop the Archimedean method for power functions of ever higher degrees, because other methods yield the same results with less effort. Nevertheless, we have seen that the method has great potential for expansion and that essential insights from modern integral calculus are already contained in it in a rudimentary form.

Bibliography

Burn, R. P. (1999): *Integration, a genetic introduction,* Nordisk matematikkdidaktikk, No. (7)1, pp. 7–27.

H. Eves, (1983), *An Introduction to the History of Mathematics*, Saunders College Publishing, 5th edition, pp. 290–291.

Kirfel, C. (2013A), *Arkimedes og parabelen*, TANGENTEN (4/2013) Tidsskrift for matematikkundervisning, pp. 46–49.

Kirfel, C. (2013B), *A generalisation of Archimedes' method, The Mathematical Gazette*, Vol. 97, No. 538 (March 2013), pp. 43–52. https://www.jstor.org/stable/24496758

Kirfel, C. (2014A), Generalisering av Arkimedes' metode, TANGENTEN (1/2014) Tidsskrift for matematikkundervisning, pp. 34–39.

Kirfel, C. (2017B), *Generalisierung der Archimedischen Methode zur Flächenbestimmung des Parabelsegments*, Beiträge zum Mathematikunterricht 2017, U. Kartenkamp & Kuzle (Hrsg.) Beiträge zum Mathematikunterricht 2017, Münster WTM-Verlag, pp. 541–544.

Toeplitz, O. (1949), *Die Entwicklung der Infinitesimalrechnung – eine Einführung in die Infinitesimalrechnung nach der genetischen Methode*, Springer Verlag.

Internet Addresses

[Archimedes] https://en.wikipedia.org/wiki/Quadrature_of_the_Parabola
[Euclid 1] http://www.claymath.org/library/historical/euclid/
[Euclid 2] http://aleph0.clarku.edu/~djoyce/java/elements/elements.html

Indian Roots

2

2.1 Where Do Roots Appear?

Extracting roots is not a simple task. It involves finding a new number that, when multiplied by itself, equals a given number. If we start with $Q = 25$ and want to find a number that multiplied by itself equals 25, the answer, of course, is 5. The question becomes more difficult when we set $Q = 2$. There is no integer that, when multiplied by itself, equals 2, because $1 \cdot 1 = 1$ is too small and $2 \cdot 2 = 4$ is too big. There are no integers between 1 and 2. Therefore, the process of extracting roots is not a trivial task.

When adding two numbers given to us as integers, or perhaps known as decimal numbers, we can quickly provide the answer using the addition algorithm, and the answer is 100% exact. The same applies to subtraction and multiplication. In division, we must be prepared for infinite decimal fractions with a repeating pattern after the decimal point, such as in $\frac{1}{3} = 0.333333\ldots$, but even here, the pattern can be found in a finite number of steps. When extracting roots—as a new kind of "calculation" so to speak—something fundamentally new occurs. Later in this chapter we will show that if you try to extract the root of a number that is not a perfect square, the digits after the decimal point will not repeat in a fixed pattern, and we will never completely finish the task of extracting the root. If the answer is to be given as a decimal fraction, the result can only be described with a certain degree of accuracy, but we never get the exact final result, assuming we are trying to extract the root of a number that is not itself a perfect square.

So far, we have looked at roots in arithmetic, or number calculation. We can also look for geometric representations of roots. For example, we might be interested in constructing a square with an area of 2 square units. The question then is, how long is the corresponding side length g of the square, because we must claim $g^2 = 2$. Since the area is calculated as length times width and in a square, both the length and width are equal to g we must have $g = \sqrt{2}$.

© The Author(s), under exclusive license to Springer-Verlag GmbH, DE, part of Springer Nature 2026

C. Kirfel, *Side Paths in the History of Mathematics*, Mathematics Study Resources 21,
https://doi.org/10.1007/978-3-662-72918-2_2

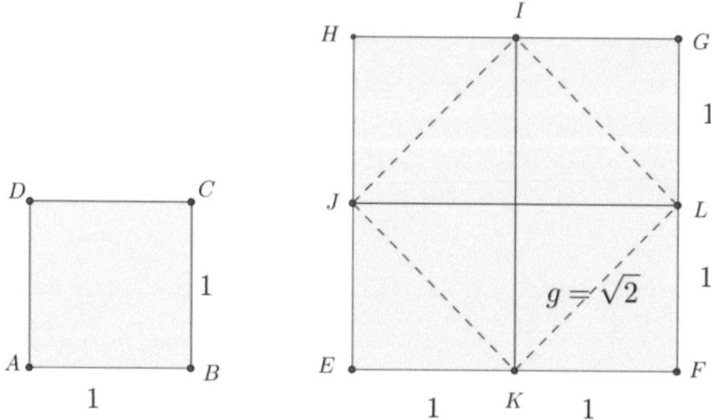

Fig. 2.1 Square root of 2

The left square *ABCD* has an area of 1, because both its length and width are 1 (see Fig. 2.1). The right square *EFGH*, on the other hand, has an area of 4, because both its length and width are 2. Therefore, the square *JKLI*, which includes exactly half of each of the four sub-squares of the large square, has an area of 2. Thus, the side length of the dashed square must be $\sqrt{2}$.

We will see how the historical writings of the ancient Vedas combine both aspects, the geometric view and arithmetic calculation, and establish new connections between these two areas of mathematics.

2.2 Historical Background

The Vedas are ancient Indian Sanskrit texts that contain religious rituals, prayers, and chants used in the Vedic Hindu religion. The Sulbasutra is a part of these Vedas, where the construction of religious altars, fireplaces called Agni, is described [Sulbasutra]. These had to be built according to mathematical rules for the offerings made there to be successful. Therefore, the Sulbasutra contains a lot of mathematics and at times reads like a geometry textbook. These texts, written by Hindu scholars around 600 BCE, contain knowledge that is probably much older (2000 BCE) and was previously transmitted orally. Henderson (2000) believes that this is the oldest geometry textbook in the world. In the Sulbasutra, for example, We find the following statement (Sutra 52 of Baudhayana's Sulbasutra): "The square on the diagonal of a square is twice as large as the square itself." (see also Fig. 2.1).

If we now give the original square a side length of 1, then the diagonal is the length of a square with an area of 2 and the diagonal itself must have the length $\sqrt{2}$. In Sanskrit, the term *dvi-karani* is used, which means "what generates the two." The Sulbasutra also contains a remarkable method for finding a numerical approximation to $\sqrt{2}$:

$$\sqrt{2} \approx \frac{3}{2} - \frac{1}{4 \cdot 3} - \frac{1}{34 \cdot 4 \cdot 3},$$

which is an impressively accurate approximation. If you write the difference of the fractions as a decimal number, five decimals after the decimal point are already correct. The technique behind this result, that Datta (1932) and Henderson (2000) present in their articles, is simple and can be easily developed further. Another step in this algorithm would already lead to the approximation

$$\sqrt{2} \approx \frac{3}{2} - \frac{1}{4 \cdot 3} - \frac{1}{34 \cdot 4 \cdot 3} - \frac{1}{1154 \cdot 34 \cdot 4 \cdot 3}.$$

If you write this as a decimal number, you already get 11 correct digits after the decimal point.

What pattern is hidden behind this old Indian algorithm, can it be generalized so that it can be used for numbers other than $\sqrt{2}$ and why does it converge so quickly, i.e., why do the calculated values approach the final result so rapidly? To the first question, we can say the following: Datta (1932) and Henderson (2000) speculate that the method might have looked like this. We start with a rectangle consisting of two unit squares. We divide one of the two into two halves and stick these onto the outside of the first square. We get an L-shaped figure.

This L-shaped figure in Fig. 2.2 has an area of 2 and it already resembles a square a bit. If it were a perfect square, its side length would be exactly $\sqrt{2}$. Thus, the side length of the L-shaped figure, 3/2, is already a useful approximation of $\sqrt{2}$. The idea now is to fill the small "missing" square with "material" from the edge of the figure and thereby construct a new L-shaped figure that resembles a square even more. The side of the new figure is then a better approximation to $\sqrt{2}$ than the 3/2 with which we started.

So, we start with the small white square with side length ½. The area is thus ¼. We now cut a strip from the left and bottom of the L-figure, whose outermost edge is $2 \cdot \frac{3}{2} = 3$ units long, and we give the strip the width $\frac{\frac{1}{4}}{3} = \frac{1}{12}$, since Area

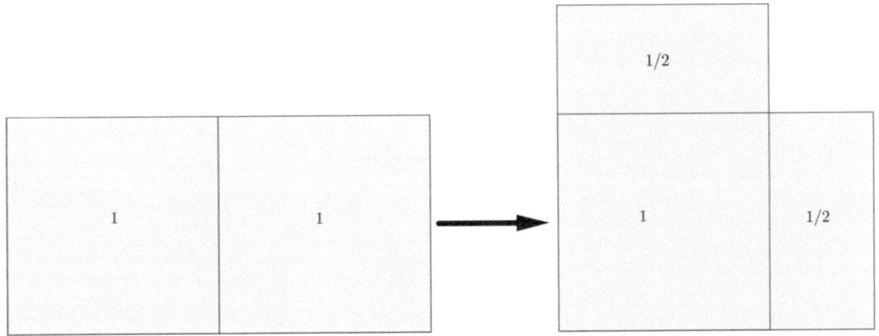

Fig. 2.2 Creating the L-figure

= Length × Width (see Fig. 2.3). We now use this strip to fill the still open white square. But, because the horizontal and vertical parts of the strip overlap, our strip is a bit too short to fill the entire white square and we get a new L-shaped figure. The overlap is again a square, this time with side length $\frac{1}{12}$. We stack the strips $A1, A2 \ldots, A6$ on top of each other to fill the square. Unfortunately, however, $A6$ is a bit too short and—as we already have mentioned—we get a new L-figure, but this time it resembles a square much more than the old one. The new side length of the L-figure $\frac{3}{2} - \frac{1}{12}$ is therefore a much better approximation to $\sqrt{2}$ than the previous one. We now hope to be able to repeat this process and produce new L-figures that resemble a square more and more, so that their side lengths are better and better approximations to $\sqrt{2}$. Here we get

$$\frac{3}{2} - \frac{1}{4 \cdot 2 \cdot \left(\frac{3}{2}\right)} = \frac{3}{2} - \frac{1}{12} = \frac{17}{12}$$

as an approximation of $\sqrt{2}$. In the next step, we need to remove a strip that has the width

$$\frac{\left(\frac{1}{12}\right)^2}{\left(2 \cdot \frac{17}{12}\right)} = \frac{1}{2 \cdot 3 \cdot 4 \cdot 17}$$

and we then need to distribute this area over the small square. The uncovered part of the figure is then a tiny square with an area of

$$\left(\frac{1}{2 \cdot 3 \cdot 4 \cdot 17}\right)^2,$$

which is less than 10^{-5}.

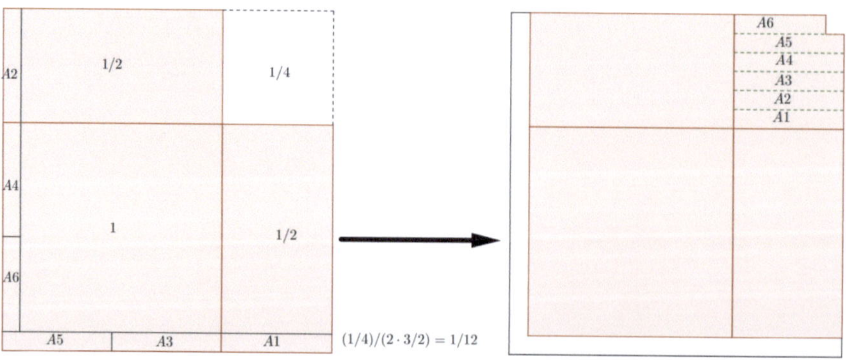

Fig. 2.3 Inserting the strip

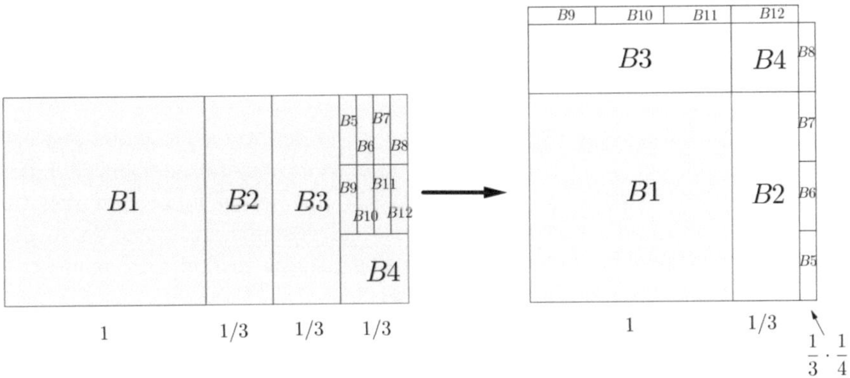

Fig. 2.4 Henderson's suggestion

Remark 2.1
The presentation here is not identical to that of Datta and Henderson. They build up the L-figure in several stages before they start "cutting". The original calculation rule from the Sulbasutra was

$$\sqrt{2} \approx 1 + \frac{1}{3} + \frac{1}{4 \cdot 3} - \frac{1}{34 \cdot 4 \cdot 3},$$

instead of $\sqrt{2} \approx \frac{3}{2} - \frac{1}{4 \cdot 3} - \frac{1}{34 \cdot 4 \cdot 3}$, which we derived above. Henderson suggests that we can arrive directly at the approximation $1 + \frac{1}{3} + \frac{1}{4 \cdot 3}$ via the following decomposition (see Fig. 2.4):

This figure is then identical to the second approximation figure in our presentation. All further approximation figures are then the same. We have chosen to start with Fig. 2.3 in order to focus on the process of cutting a strip and filling a square right from the beginning even if we are not completely in line with the historical reconstruction.

In this chapter we will consider a number of questions.

1. Can the algorithm be generalized and calculate roots of arbitrary numbers and how can it be described algebraically (with formulas)?
2. Can we always find a suitable starting figure?
3. How quickly does the process converge?
4. How can we achieve the desired accuracy?
5. Can the algorithm be compared with other known algorithms?
6. Can we make statements about the irrationality of roots?
7. Can the algorithm be linked with Newton's method?
8. Are there connections to Pell's equation?

2.3 Is There a General Formula Representation for This Process that Works for the Square Root of Any Number?

We now leave the specific example $\sqrt{2}$ and try to develop a procedure that calculates any root, i.e., $\sqrt{N}$. First, we want to consider natural numbers N. Later, we will see how fractions can be handled, too. The initial figure describes the difference area between two squares, whose side lengths are a and b, so that $N = a^2 - b^2$. We will postpone the question of how to find such numbers a and b until Sect. 2.6 "Starting Figures" and Sect. 2.7 "Other Starting Figures". So, we start with an L-shaped figure, which initially consists of a large square a^2, from which a small square b^2 is removed in the upper right corner. The area b^2 of the missing square is to be filled with "material" that we find along the bottom and left edge of the L-figure. This edge has length $2 \cdot a$. A strip of this length must therefore have width

$$b' = \frac{b^2}{2 \cdot a},$$

if it is to "match" the area of the small square b^2. However, the strip has a kink, and a part of the horizontal strip overlaps with a part of the vertical strip. Therefore, we cannot fill the entire empty space of the small square b^2. But at least we can fill a large part of the open square (see Fig. 2.5). Here we need not think of the area as a number of strips that we have to paste into the "hole" but we can imagine the area as a kind of dough that is rolled out and fills the major part of the missing square. The remaining part is a new smaller square, where the length and width are equal to the strip width $b' = b^2/(2a)$.

The new figure thus again has an L-shape and the dimensions of the new squares are

$$a' = a - b' = a - \frac{b^2}{2 \cdot a} \text{ and } b' = \frac{b^2}{2 \cdot a}.$$

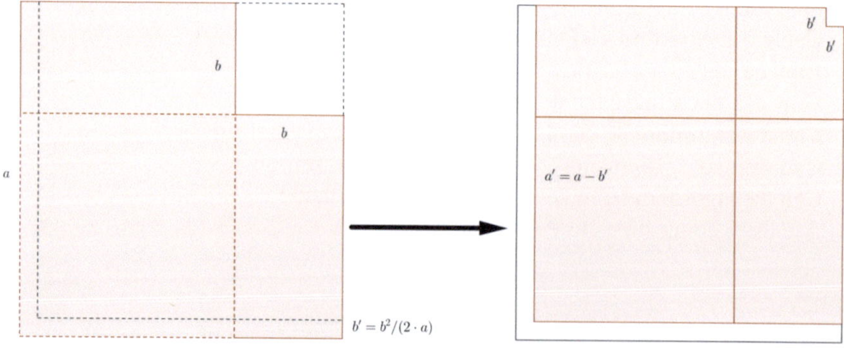

Fig. 2.5 Inserting the strip in the general case

With this, we have an algebraical description of the transition and can now use these formulas again and again. At the same time the formulas make sure that the new dimensions of the squares are rational numbers in order to produce rational approximations to the roots we look for. These formulas will be crucial for the remainder of the chapter.

2.4 Henderson's Concerns

Notice how the strips in Fig. 2.3 fit exactly into the small square, that is that the width of the strips is a divisor of $a - b$. Henderson (ibid.) writes, "This method will work for any number $N = a^2 - b^2$, where the side a is an integral multiple of the side b." He is therefore of the opinion that a must be an integral multiple of b for the process to work. Above, we saw how the 6 strips of thickness 1/12 fit into the missing square and it gives the impression that the thickness of the strips must be a divisor of the small square side. This is correct as long as we imagine that the strip is cut out of a physical material, e.g., paper, and glued into the small square layer by layer. The width of the strip $b' = \frac{b^2}{2 \cdot a}$ must then be a divisor of the length of the small square side b in this representation, and therefore

$$\frac{b}{\left(\frac{b^2}{2a}\right)} = \frac{2a}{b}$$

must be an integer, which is fulfilled as soon as b is a divisor of a. Thus, the Indian method would only be applicable for a very limited set of numbers, namely only those numbers $N = a^2 - b^2$, where a is divisible by b. But is this actually necessary? As mentioned above we can also imagine the strip as if it were made of a kind of dough, which is clumped together and then rolled out again. It is important that the area of the strip is equal to that of the small square minus the tiny new corner square. This allows the method to be carried out with all L-figures. It is therefore much more general than Henderson believes.

2.5 Geometric Considerations

Figure 2.6 shows two squares with side lengths $a = 4$ and $b = 3$, where the difference of the squares is $N = a^2 - b^2 = 7$. The L-shaped figure appears relatively narrow, raising the question of whether the area transfer (cutting and pasting the strip) as in our previous example is even possible here. We examine the situation more closely and find that

$$\frac{9}{8} = b' = \frac{b^2}{2 \cdot a} > a - b = 1$$

The dashed lines mark the strip that should be cut away. However, we do not have that much area available. We call this situation the "thin L-figure". Here, our geometric intuition of a strip that needs to be cut away fails. Nevertheless, we can perform an area transfer. We imagine that we let the given area of 7 square units flow into the dashed square as an L-shape. The dashed square has a side length of $4-\frac{9}{8}=\frac{23}{8}$. Its area $\left(\frac{23}{8}\right)^2=\frac{529}{64}\approx 8.26>7$ is larger than the colored initial area, so the mentioned "flow" of the initial area into the dashed square is possible. In general, we have for the dashed square:

$$(a')^2=(a-b')^2=a^2-2ab'+b'^2=a^2-b^2+b'^2>a^2-b^2=N,$$

so that the initial area N can easily flow into the dashed square $a'\cdot a'$ because the square is larger. We thus obtain again an L-shaped figure, the difference of two squares, now with the dimensions a' and b', where again $a'^2-b'^2=\left(a-\frac{b^2}{2a}\right)^2-\left(\frac{b^2}{2a}\right)^2=a^2-b^2=N$ applies. The area is therefore the same as in the initial figure. From the dashed square $\left(a-\frac{b^2}{2a}\right)^2$ we have to subtract the square $\left(\frac{b^2}{2a}\right)^2$ to get a figure with the same area as the original figure. At that point we left the geometric "puzzle game" of clipping and pasting strips and used algebraic arguments to be sure that the area of the new L-figure is the same as the previous one. The transition $a\to a'$ and $b\to b'$ can simply be seen as the transformation of one L-figure into another L-figure of the same area, without necessarily having to view this as cutting and pasting a strip. Here, we can rather imagine the flow of the given area into the dashed square to form a new L-figure in order to make the "missing" square smaller than the previous one. At the same time our transition formulas guarantee that the new approximation a' again is a rational number.

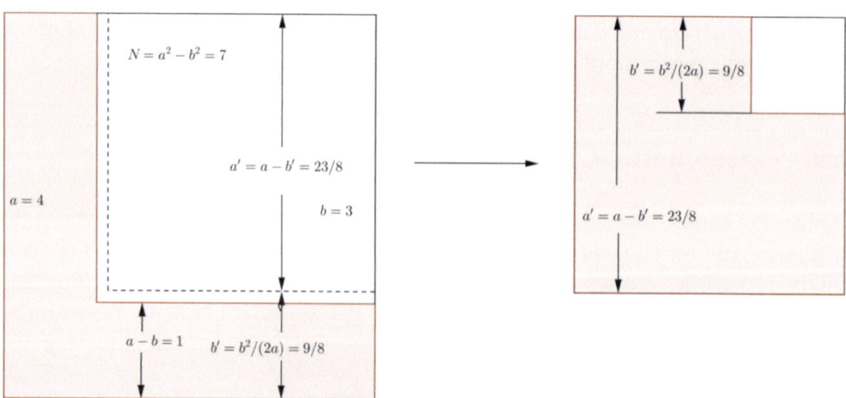

Fig. 2.6 Area transfer

2.6 Starting Figures

With the formula $N = a^2 - b^2 = (a + b)(a - b)$ as a starting point, we can immediately find an L-figure for every N if we set $N = a + b$ and $a - b = 1$. If we solve the associated system of equations

$$a + b = N$$

$$a - b = 1,$$

we get $a = \frac{N+1}{2}$ and $b = \frac{N-1}{2}$. This immediately gives us a suitable, but often—especially for large N values—very narrow starting figure. For even N, the values for a and b are not integers. If N is also divisible by 4, we can write $N/2 = a+b$ and $a - b = 2$, where of course $N = a^2 - b^2 = (a + b)(a - b)$ still applies. Then $a = N/4+1$ and $b = N/4-1$ apply, and we again get integers a and b, but where the L-figure is twice as thick (2 length units) as the previous L-figure.

If we know a divisor c of N, we can argue in the same way: We set $N/c = a+b$ and $a - b = c$, where of course $N = a^2 - b^2 = (a + b)(a - b)$ still applies. Then $a = N/(2c) + c/2$ and $b = N/(2c) - c/2$ apply, and we again have an L-figure. This time the "thickness" of the L-strip, $a - b = c$, is significantly larger than in the initial figure, where $a - b = 1$.

In any case the formula $N = a^2 - b^2 = (a + b)(a - b)$ gives us a possible L-shape as a starting figure, even though the L-shape may be very thin. For large values of N however, we always get thin L-figures and it may be worthwhile to look for alternative starting figures, which will be discussed in detail in the next section.

2.7 Other Starting Figures

Next, we would like to describe a "shortcut" that we can take to quickly get a good starting figure, where the geometric intuition of cutting and pasting a strip can be used. We continue to consider the example $N = a^2 - b^2 = 7$, but do not set $a = 4$ and $b = 3$. Instead, we look for values that lead us to a "better" starting figure. We use a trick for this.

This time we start with a square with 4 area units (the largest square number below 7) and paste the remaining area of 3 units onto the sides of the square. This gives us an L-shape, where the squares have the dimensions $a = 2 + \frac{3}{4} = \frac{11}{4}$ and $b = \frac{3}{4}$ (see Fig. 2.7). Here is $\frac{b^2}{2 \cdot a} < a - b$ and we need not think about the geometrical concerns mentioned in Sect. 2.5.

In general, it is worth starting with the largest square number n^2 strictly below N. However, for large numbers N it is not trivial to find such a number. This is a difficult task in itself. In such cases, it may then be necessary to instead set $a = \frac{N+1}{2}$ and $b = \frac{N-1}{2}$.

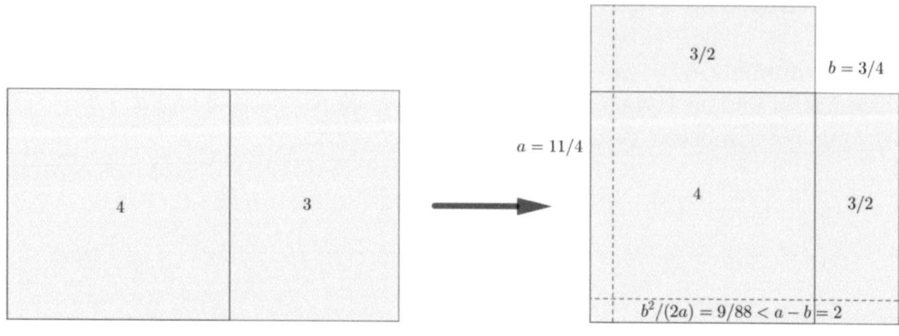

Fig. 2.7 New starting figure

If, however, we know such a number n, we can construct a geometric figure consisting of the square and a rectangle representing the remaining area $R = N - n^2 > 0$. We distribute this remaining area over two rectangles, which we attach to the sides of the square. Then we have

$$n^2 < N < (n+1)^2 = n^2 + 2n + 1$$

$$0 < R = N - n^2 < 2n + 1.$$

Since each of the rectangles now has an area of $R/2$ area units and the width n, the length must be $b = \frac{R}{2n}$ and thus $a = n + \frac{R}{2n}$. Because R and n are natural numbers, with $R < 2n + 1$ we also have $R \leq 2n$ or $\frac{R}{2n} \leq 1$, and therefore

$$\frac{b^2}{2a} = \frac{\left(\frac{R}{2n}\right)^2}{2\left(n + \frac{R}{2n}\right)} < 1/2$$

and thus $a - b = n \geq 1 > \frac{b^2}{2 \cdot a}$. This means that we do not have the problem of a thin L-figure and we can continue to use our geometric intuition of cutting material from the edge and let it flow into the missing square.

Remark 2.2
In the book "The crest of the peacock" (Joseph 1990, pp. 260 and 261) the Bakhshali manuscript is mentioned. The manuscript contains a method that reminds us of the first steps of the algorithm from the Sulbasutra. Here it is specifically mentioned that the next square number below N is a good starting point.

2.8 Examples

At this point, it is worth calculating a few examples to see what can occur during the calculation steps. In all examples we choose start with $a = \frac{N+1}{2}$ and $b = \frac{N-1}{2}$.

Example 1: $\sqrt{2}$

Here, as we already know, we can start with $a = \frac{N+1}{2} = 3/2$ and $b = \frac{N-1}{2} = 1/2$. Then $a' = 17/12$ and $b' = 1/12$. With the formulas $b' = b^2/(2a)$ and $a' = a - b'$ we then get Table 2.1. Starting from the fourth row in the table, the approximation values for $\sqrt{2}$ no longer change, which is due to the fact that we already have found so many correct digits. For practical use, it is often sufficient to perform only three or four steps of the algorithm.

Example 2: $\sqrt{19}$

Here we start with $a = \frac{N+1}{2} = 10$ and $b = \frac{N-1}{2} = 9$. Then $a' = 119/20$ and $b' = 81/20$. Using the formulas $b' = b^2/(2a)$ and $a' = a - b'$ we then obtain Table 2.2. We see that the process does not converge as quickly as in the $\sqrt{2}$ example, and that we have only reached three correct decimals even after 4 steps.

Example 3: $\sqrt{111}$

Table 2.1 Numerical calculation of $\sqrt{2}$

a-values numerator	a-values denominator	b-values numerator	b-values denominator	a-values	$\sqrt{2}$
3	2	1	2	1.5	
17	12	1	12	1.416666667	1.41421356237
577	408	1	408	1.414215686	
665,857	470,832	1	470,832	1.414213562	
8.87E + 11	6.27E + 11	1	6.27E + 11	1.414213562	

Table 2.2 Numerical calculation of $\sqrt{19}$

a-values Numerator	a-values Denominator	b-values Numerator	b-values Denominator	a-values	$\sqrt{19}$
10	1	9	1	10	
119	20	81	20	5.9500000000	4.35889894354
21,761	4760	6561	4760	4.57163865546	
9.04E + 08	2.07E + 08	43,046,721	2.07E + 08	4.36384883005	
1.63E + 18	3.75E + 17	1.85E + 15	3.75E + 17	4.35890175085	

Table 2.3 Numerical calculation of $\sqrt{111}$

a-values Numerator	a-values Denominator	b-values Numerator	b-values Denominator	a-values	$\sqrt{111}$
56	1	55	1	56	
3247	112	3025	112	28.99107143	10.53565375285
11,935,393	727,328	9,150,625	727,328	16.40991822	
2.01E + 14	1.74E + 13	8.37E + 13	1.74E + 13	11.58706006	
7.39E + 28	6.99E + 27	7.01E + 27	6.99E + 27	10.5833559	

Here we start with $a = \frac{N+1}{2} = 56$ and $b = \frac{N-1}{2} = 55$. Then $a' = 3247/20$ and $b' = 3025/110$. With the formulas $b' = b^2/(2a)$ and $a' = a - b'$ we then obtain Table 2.3. We see that even after four steps in the algorithm, we only have one correct decimal. This is not only due to the magnitude of the number N but also due to our choice of starting values $a = \frac{N+1}{2}$ and $b = \frac{N-1}{2}$. It is therefore worthwhile to consider the so-called convergence speed, which we want to do in the next section.

2.9 How Fast Does the Process Converge?

In our L-figure we always have $a > b$, and thus we automatically get $b' = \frac{b^2}{2a} < \frac{b}{2}$, which means that the side length of the free square is at least halved at each step, making the b-sequence a null sequence and convergence is guaranteed.

Now we want to examine the effectiveness of the algorithm. As we have seen, the sequence of the a values, i.e., $a, a', a'', \ldots$ are estimates of $\sqrt{N}$. We now want to say something about the distance of the approximation to the target $\sqrt{N}$ and especially examine how this distance decreases over the course of the iterations, i.e., we want to find a relationship between $a - \sqrt{N}$ and $a' - \sqrt{N}$. We use the term "iteration" here to refer to the repeated calculations of the values a and b. This relationship will then tell us how much better the new approximation $a' - \sqrt{N}$ is compared to the old $a - \sqrt{N}$.

We have $N = a^2 - b^2$ but also $N = a'^2 - b'^2$, because the two L-shaped figures in Fig. 2.5 have the same area. We therefore have

$$a'^2 - N = b'^2$$

$$\left(a' - \sqrt{N}\right)\left(a' + \sqrt{N}\right) = b'^2$$

$$a' - \sqrt{N} = \frac{b'^2}{a' + \sqrt{N}} = \frac{\left(\frac{b^2}{2a}\right)^2}{a' + \sqrt{N}} = \frac{b^4}{(2a)^2\left(a' + \sqrt{N}\right)}$$

$$= \frac{\left(a - \sqrt{N}\right)^2}{a' + \sqrt{N}} \cdot \left(\frac{a + \sqrt{N}}{2a}\right)^2$$

$$< \frac{\left(a - \sqrt{N}\right)^2}{2\sqrt{N}} < \frac{\left(a - \sqrt{N}\right)^2}{2}.$$

The last fraction before the first inequality sign is < 1, because $\sqrt{N} < a$. The main result is thus

$$a' - \sqrt{N} < \frac{\left(a - \sqrt{N}\right)^2}{2}. \tag{2.1}$$

This is called quadratic convergence and is a very good result. If the original approximation is less than 1/1000, the subsequent approximation is already less than 1/2,000,000. With each step, the number of correct decimals will at least double. If you start with 1.5 as an approximation to $\sqrt{2}$, which is less than 1/10 away from the correct answer, then after one iteration—i.e., after one step with new calculations of a and b—you will be at most 1/200, after 2 iterations at most 1/80,000, and after three iterations at most 10^{-10} away from the correct answer, which means that we have at least 10 correct decimals.

We will now assume that our initial figure, which we described in the Sect. 2.7 "Other Starting Figures", is the L-figure with dimensions $a = n + R/(2n)$ and $b = R/(2n)$, where n is the largest integer whose square is below N and $R = N - n^2$. We now have

$$a^2 - N = b^2$$

$$\left(a - \sqrt{N}\right)\left(a + \sqrt{N}\right) = b^2$$

$$a - \sqrt{N} = \frac{R^2}{4n^2\left(a + \sqrt{N}\right)} \leq \frac{1}{2 \cdot \sqrt{N}} \leq \frac{1}{2}.$$

Here we have used that $R \leq 2n$ and that $a > \sqrt{N}$. This result means that the distance of the first approximation from $\sqrt{N}$ is already less than ½. So, the distance becomes smaller and smaller with each further step through squaring and halving (2.1) and finally converges to zero. Our intuition supports this conclusion, since we fill in parts of the missing square in each step. Thus, the approximation to the final square must get better with each step. We observe that by starting with the largest square number below N, we will be squaring and halving a quantity that was initially less than 1, so we double the number of correct digits right from the first iteration.

In Example 1 above, we saw that the first approximation $a = 3/2$ was already less than a tenth away from the correct answer $\sqrt{2}$. With each step in the algorithm, the number of correct decimals doubles, as we can see in Table 2.1.

In the second example with $\sqrt{19}$, this is also the case, namely that the number of correct decimals doubles at each step once we have one correct decimal after the decimal point. However, this occurs a bit later here.

In the third example with $\sqrt{111}$, it takes even longer until we have the first correct decimal. After that, the number of correct decimals doubles at each step.

Since we have developed formulas for both the calculations of a' and b' and formulas for the starting values $a = (N+1)/2$ and $b = (N-1)/2$, the algorithm is suitable for programming. The next section will examine how a desired accuracy can be obtained.

2.10 How Can a Desired Accuracy Be Achieved?

Above, we saw that

$$a' - \sqrt{N} < \frac{\left(a - \sqrt{N}\right)^2}{2}.$$

After the next step, we have

$$a'' - \sqrt{N} < \frac{\left(a' - \sqrt{N}\right)^2}{2} < \frac{\left(a - \sqrt{N}\right)^4}{4}.$$

If you perform several steps, you get similar results. To avoid symbols with many apostrophes, we introduce a new notation. We set:

$$a = a_0, a' = a_1, a'' = a_2, \ \ldots \text{ and } b = b_0, b' = b_1, b'' = b_2 \ldots$$

Now let's assume a desired accuracy is given, e.g., a root should be determined accurately to eight decimals. This means we need to determine the number of steps k that must be run through so that

$$a_k - \sqrt{N} < \frac{1}{10^8}.$$

Following the above pattern (2.1), then $a_k - \sqrt{N} < \frac{\left(a_0 - \sqrt{N}\right)^{2k}}{2^k}$. If it is known that the first approximation is $a_0 - \sqrt{N} < 1/10$, we set

$$a_k - \sqrt{N} < \frac{\left(a_0 - \sqrt{N}\right)^{2k}}{2^k} < \frac{\left(\frac{1}{10}\right)^{2k}}{2^k} = \frac{1}{200^k} < \frac{1}{10^8} \quad \text{i.e. } 10^8 < 200^k,$$

so $3<k<4$ and four calculation steps are sufficient to achieve the desired accuracy.

In the general case, where the initial accuracy is unknown, we can simply perform the calculation steps of the algorithm until $b_k^2 < 1/5$. Then

$$a_k - \sqrt{N} = \frac{b_k^2}{a_k + \sqrt{N}} < \frac{b_k^2}{2\sqrt{N}} < \frac{b_k^2}{2} < \frac{1}{10}.$$

Then the k-th approximation is less than $1/10$ away from the root and we have at least one correct decimal. From there, we can be sure that the number of correct decimals doubles at each step, and thus we can calculate the number of remaining steps that lead to the desired accuracy.

Remark 2.3

Instead of the estimate $a' - \sqrt{N} < \frac{\left(a-\sqrt{N}\right)^2}{2}$, we could have used the much better estimate $a' - \sqrt{N} < \frac{\left(a-\sqrt{N}\right)^2}{2\sqrt{N}}$, which also appears above. Here you can see that the desired accuracy is achieved even faster, especially when N is large.

2.11 Can the Algorithm Be Compared with Other Known Algorithms?

A common algorithm for calculating square roots is Heron's method. In English-speaking countries, this method is often called the "divide and average" method. To calculate $\sqrt{N}$, we start with a starting value, e.g., $s = 1$, and then calculate new approximations to $\sqrt{N}$ using the following "divide and average" formula

$$s' = \frac{\frac{N}{s} + s}{2}$$

So, we divide N by the approximation s and then take the average of this result and the approximation s, i.e., "divide and average". The new value s' is then reused as the starting value in the same formula. In this way, we constantly produce better approximations to $\sqrt{N}$. For example, if we start with $s = 1$ as an "approximation" to $\sqrt{2}$, this gives us $s' = \frac{\frac{2}{1}+1}{2} = 3/2$ as the next approximation to $\sqrt{2}$, and $s'' = \frac{\frac{2}{3/2}+\frac{3}{2}}{2} = \frac{17}{12}$ is an even better approximation to $\sqrt{2}$. The numerical values $s = \frac{3}{2}$ and $s' = \frac{17}{12}$ may seem familiar. They already appeared in the Indian algorithm, and it raises the suspicion that the two algorithms are related. This is indeed the case.

Let us imagine that the Indian algorithm has already reached an approximation a in the calculation of $\sqrt{N}$, where $N = a^2 - b^2$. We now use Heron's formula for

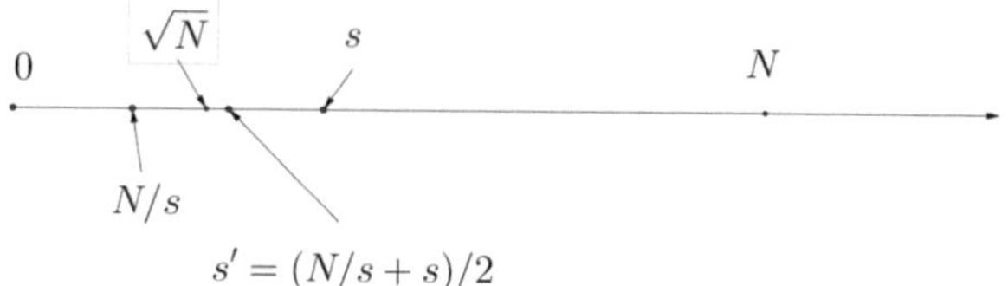

Fig. 2.8 Heron's method

this approximation value a and obtain

$$a^* = \frac{\frac{N}{a} + a}{2} = \frac{\frac{a^2-b^2}{a} + a}{2} = \frac{a - \frac{b^2}{a} + a}{2} = a - \frac{b^2}{2a}$$

which is the well-known formula from the Indian algorithm. That is, Heron's formula and the Indian algorithm produce the same sequences of numbers. They are therefore equivalent.

If we start the process of Heron's method with $s = 1$, the next approximation value is $s' = \frac{\frac{N}{1}+1}{2} = \frac{N+1}{2}$. This corresponds exactly to the starting figure, where $N = a + b$ and $1 = a - b$ applies, which we have discussed in detail above.

Heron's method can be illustrated on the number line (see Fig. 2.8). The goal is to calculate $\sqrt{N}$. We start with an approximation value s, then N/s is also an approximation to $\sqrt{N}$, and it makes sense to use the average of these two approximation values as the next approximation.

Since the arithmetic mean $\left(\frac{N}{s} + s\right)/2$ of the numbers N/s and s always is larger than the geometric mean $\sqrt{\left(\frac{N}{s}\right)s} = \sqrt{N}$ we see that our approximations always are larger than $\sqrt{N}$ when we use Heron's formula.

It is astonishing that Heron's method was already anticipated in the Indian method of root extraction about 600 BCE. On the one hand, the Indian method is immediately clear due to the geometric representation, and on the other hand, it provides a high speed of approximation, both of which are properties that are highly valued in mathematics.

2.12 Newton's Method

Newton developed a general method for approximating the zeroes of functions. If we set $f(x) = x^2 - N$, the zeroes of f are just $\pm\sqrt{N}$. To extract the root of N, we can therefore search for a zero of f. Newton's method starts with a starting value x_n and the corresponding point $A = (x_n, f(x_n)) = \left(x_n, x_n^2 - N\right)$ on the graph of the function. We draw the tangent at this point, which has the equation

$$\frac{y - \left(x_n^2 - N\right)}{x - x_n} = f'(x_n) = 2x_n \ i.e. \ y = 2x_n(x - x_n) - N + x_n^2 = 2x_n x - x_n^2 - N.$$

We now want to find the intersection of the tangent and the x-axis, so we set $y = 0$, which gives us $N + x_n^2 = 2x_n x$ or $x = \left(N + x_n^2\right)/(2x_n)$. This intersection is then close to the root of the function, usually much closer than the initial value (see Fig. 2.9). This zero is then the next approximation value x_{n+1} and we have

$$x_{n+1} = \frac{N + x_n^2}{2x_n} = \frac{\frac{N}{x_n} + x_n}{2}.$$

We see, therefore, that Newton's method and Heron's method, and thus also the Indian algorithm, coincide.

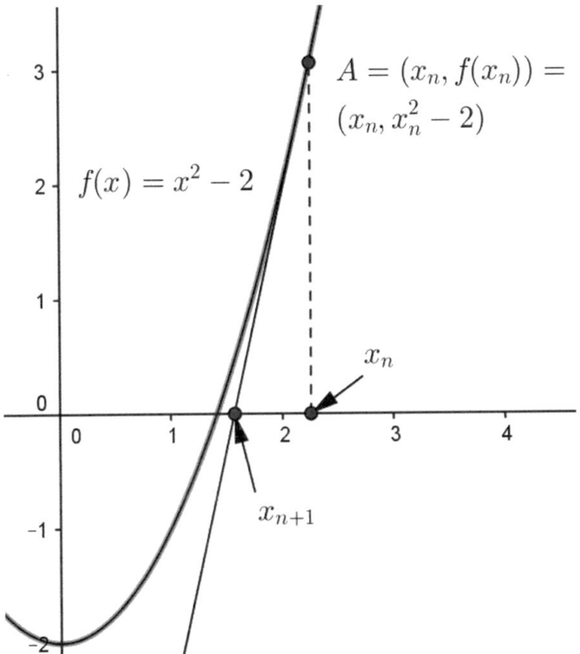

Fig. 2.9 Newton's method

2.13 Root Extraction with the Taylor Series

Taylor series can also be used to calculate roots. The following applies:

$$\sqrt{1+x} = (1+x)^{1/2} = 1 + \frac{x}{2} - \frac{x^2}{8} + \frac{x^3}{16} - \cdots$$

This works particularly well when x is very small. Further below (Remark 2.5), we will see that we can calculate $\sqrt{1.7}$ instead of $\sqrt{170}$, two numbers that consist of the same sequence of digits and only differ in that the comma is in different places. This gives us the opportunity to easily move the number we want to extract the root of close to 1. We can then use the above formula for the Taylor series and expect convergence. However, it turns out that the Taylor series converges much slower than the Indian method. With each extension of the sum with a new term, far fewer new correct digits are obtained than by the Indian method. Therefore, we will not delve into this method here.

Remark 2.4
In the 1960s, many schools still taught the handwritten algorithm for root extraction, which is very similar to the division algorithm. In each step, one new digit was found after the decimal point. Such an algorithm, of course, cannot be compared to the present Indian algorithm, where the number of correct digits at least doubles in each step. Thus, the Indian method easily outperforms the handwritten algorithm.

2.14 Irrationality of Roots

There is no indication that the Indians who wrote the Sulbasutra were concerned with the problem of the irrationality of numbers. They were aware that in the mentioned constructions, a tiny square was always missing for the exact solution of the problem, but whether the desired final solution could no longer be expressed as a fraction apparently did not interest them. It is all the more astonishing that it is nevertheless possible to show with the methods developed by the Indians that roots of numbers that are not themselves perfect squares are irrational numbers.

It is the structure of the sequence that the Indian algorithm for roots from integers produces, especially the phenomenon of nested denominators, that is suitable for demonstrating the irrationality of these roots. Here we refer to the proof of irrationality for the Euler number e by Joseph Fourier (1815).

2.15 Irrationality of the Euler Number e

The irrationality of e was first proved by Leonhard Euler in 1737. However, we will follow in the footsteps of Joseph Fourier from 1815 [Fourier] and use the following definition:

$$e = 1 + \frac{1}{1!} + \frac{1}{2!} + \frac{1}{3!} + \frac{1}{4!} + \cdots = \sum_{i=0}^{\infty} \frac{1}{i!}.$$

Suppose e could be written as a fraction, i.e., $e = p/q$, we would get

$$\frac{p}{q} = e = 1 + \frac{1}{1!} + \frac{1}{2!} + \frac{1}{3!} + \frac{1}{4!} + \cdots + \frac{1}{n!} + \cdots$$

If we multiply both sides by $q!$ we get

$$\begin{aligned}(q-1)!p &= q!\left(1 + \frac{1}{1!} + \frac{1}{2!} + \frac{1}{3!} + \frac{1}{4!} + \cdots + \frac{1}{n!} + \cdots\right)\\ &= \left(q! + \frac{q!}{1!} + \frac{q!}{2!} + \frac{q!}{3!} + \frac{q!}{4!} + \cdots + \frac{q!}{q!}\right)\\ &\quad + \frac{q!}{(q+1)!} + \frac{q!}{(q+2)!} + \frac{q!}{(q+3)!} + \cdots\end{aligned}$$

or

$$\begin{aligned}&(q-1)!p - \left(q! + \frac{q!}{1!} + \frac{q!}{2!} + \frac{q!}{3!} + \frac{q!}{4!} + \cdots + \frac{q!}{q!}\right)\\ &= \frac{q!}{(q+1)!} + \frac{q!}{(q+2)!} + \frac{q!}{(q+3)!} + \cdots\\ &< \frac{1}{2} + \frac{1}{2\cdot 2} + \frac{1}{2\cdot 2\cdot 2} + \cdots\\ &= 1\end{aligned}$$

The left side, being the difference of integers, is naturally an integer, while the right side lies between zero and one. This is impossible, and our assumption that e can be written as a fraction is contradicted. In the last inequality, we used the sum for an infinite geometric series.

We now use the idea underlying the above result to formulate a somewhat more general theorem.

Theorem 1 Let $c_i, i \geq 1$ be a sequence of natural numbers with $c_i \geq 2$ and $\lim_{i\to\infty} c_i = \infty$. We set $q_i = c_1 \cdot c_2 \cdot \cdots \cdot c_i$. Then the number $\alpha = \sum_{i=1}^{\infty} \frac{1}{q_i}$ is irrational. We say that the denominators q_n have a nested structure because each denominator is completely contained as a factor in the subsequent denominator.

Proof. Because $q_i = c_1 \cdot c_2 \cdot \dots \cdot c_i \geq 2^i$ we have $0 < \alpha = \sum_{i=1}^{\infty} \frac{1}{q_i} \leq \sum_{i=1}^{\infty} \frac{1}{2^i} = 1$ and we know that α exists as a real number. Suppose $\alpha = p/q$ could be written as a fraction, then we would have $\frac{p}{q} = \sum_{i=1}^{\infty} \frac{1}{q_i}$. We now choose n so large that $c_{n+1} > 2q$ and multiply both sides of the equation by $q \cdot q_n$. Then we get

$$p \cdot q_n = q \cdot q_n \sum_{i=1}^{\infty} \frac{1}{q_i} \text{ or } p \cdot q_n = q \cdot q_n \left(\frac{1}{q_1} + \frac{1}{q_2} + \frac{1}{q_3} + \cdots \frac{1}{q_n} \right) + q \cdot q_n \sum_{i=n+1}^{\infty} \frac{1}{q_i}.$$

or

$$\begin{aligned} p \cdot q_n - q \cdot \left(\frac{q_n}{q_1} + \frac{q_n}{q_2} + \frac{q_n}{q_3} + \cdots \frac{q_n}{q_n} \right) &= q \sum_{i=n+1}^{\infty} \frac{q_n}{q_i} \\ &= q \left(\frac{1}{c_{n+1}} + \frac{1}{c_{n+1}c_{n+2}} + \frac{1}{c_{n+1}c_{n+2}c_{n+3}} + \cdots \right) \\ &\leq \frac{q}{c_{n+1}} \left(1 + \frac{1}{2} + \frac{1}{2 \cdot 2} + \frac{1}{2 \cdot 2 \cdot 2} + \cdots \right) < 1 \end{aligned}$$

So, we have obtained the same contradiction as above. The left side is an integer, while the right side lies between zero and one. We must therefore reject our assumption that α can be written as a fraction.

Remark 2.5

In a geometric series such as $1 + \frac{1}{2} + \frac{1}{4} + \frac{1}{8} + \cdots \frac{1}{2^n} + \cdots = 2$ there is also a nested denominator structure because each denominator is fully contained as a factor in the subsequent denominator. However, the condition $\lim_{i \to \infty} a_i = \infty$ is not fulfilled because $c_i = 2$ for all i. Here we cannot apply Theorem 1 and here the sum (the number 2) is rational. In a way, the number e is the simplest number that fulfills the condition of nested denominators and increasing c_i since $c_i = i \geq 2$ from the third term in the sum. In addition, the c_i are also integers.

We now return to the approximations given by the Indian algorithm and first consider

$$\sqrt{2} \approx \frac{3}{2} - \frac{1}{3 \cdot 4} - \frac{1}{3 \cdot 4 \cdot 17} - \frac{1}{3 \cdot 4 \cdot 17 \cdot 1{,}154}$$

We see that this is a sum of unit fractions, except for the first term, where there is probably a nested denominator structure, and we can conclude using Theorem 1 that $\sqrt{2}$ is irrational.

For $\sqrt{2}$ we can start with $a = \frac{p}{q} = \frac{3}{2}$, $b = \frac{v}{q} = \frac{1}{2}$ and then have $b' = \frac{b^2}{2a} = \frac{v^2}{2pq} = \frac{1}{12}$, $a' = a - b' = \frac{p}{q} - \frac{v^2}{2pq} = \frac{2p^2 - v^2}{2pq} = \frac{17}{12}$. Then we see that the denominators of the b-terms have a nested structure. The denominator q of b is completely contained as a factor in the denominator $2pq$ of b'. In the next step, the denominator $2pq$ of b' is completely contained in the denominator $2pq2(2p^2 - v^2)$ of b'' and so on. This is of course because we, in order to find a new approximation, have to subtract one fraction from another. Here we have to use the main denominator, which is essentially the product of the individual denominators, and thus guarantees us the nested denominator structure. We now want to use this to create a connection between the Indian root algorithm and irrational quantities, i.e., Theorem 1. Because the a-values decrease at each step by the new b-value, we get

$$\sqrt{N} = a_0 - b_1 - b_2 - \cdots = a_0 - \sum_{i=1}^{\infty} b_i.$$

Thus, we obtain a representation of $\sqrt{N}$ as an infinite sum, which fits into our Theorem 1. We can now apply Theorem 1 to our series for $\sqrt{2}$. Here are the details: We set $M_1 = p = 3, M_2 = 2p^2 - 1 = 17, M_3 = 2\left(2p^2 - 1\right)^2 - 1 = 577$ and $M_{i+1} = 2M_i^2 - 1$. Then $c_i = 2M_i$ and the sequence grows to infinity. Therefore, with $b_i = \frac{1}{q_i}$ we have that

$$\sqrt{2} = \frac{3}{2} - b_1 - b_2 - \cdots = \frac{3}{2} - \sum_{i=1}^{\infty} \frac{1}{q_i}$$

is an irrational number.

The case is very similar when you want to calculate $\sqrt{15}$. Here you can start with $a = 4$ and $b = 1$ and get $p = 4, q = v = 1$. Thus $M_1 = p = 4, M_2 = 2p^2 - 1 = 31, M_3 = 2\left(2p^2 - 1\right)^2 - 1 = 1,921$ and $M_{i+1} = 2M_i^2 - 1$. Then $c_i = 2M_i$ an grows to infinity. Therefore we have that

$$\sqrt{15} = 4 - b_1 - b_2 - \cdots = 4 - \sum_{i=1}^{\infty} \frac{1}{q_i}$$

is irrational.

The following example, however, shows that Theorem 1 cannot help us in all cases.

We examine $\sqrt{19}$ and place 19 between two consecutive square numbers $4^2 = 16 < 19 < 25 = 5^2$. That is, we can start with an L-figure that has the dimensions $a = 4 + \frac{3}{8} = \frac{35}{8}$ and $b = \frac{3}{8}$. We have $a^2 - b^2 = \left(\frac{35}{8}\right)^2 - \left(\frac{3}{8}\right)^2 = \frac{1{,}225-9}{64} = \frac{1{,}216}{64} = 19 = N$. Here we have

$$\sqrt{19} \approx \frac{35}{8} - \frac{3^2}{2 \cdot 8 \cdot 35} - \frac{3^4}{2 \cdot 8 \cdot 35 \cdot 2 \cdot 2{,}441} - \cdots$$

Again, we see that there is a nested denominator structure. But this time we do not get unit fractions. On the contrary, the numerators form a rapidly growing sequence. So, we have to generalize Theorem 1 to also include numerators $\neq 1$. This is the objective of Theorem 2.

Theorem 2 Let $c_i, i \geq 1$ and $t_i, i \geq 1$ be two sequences of natural numbers with $c_i \geq 2$,so that $\lim_{i\to\infty} \frac{t_i}{c_i} = 0$. We again set $q_i = c_1 \cdot c_2 \cdot \dots \cdot c_i$. Then the number $\alpha = \sum_{i=1}^{\infty} \frac{t_i}{q_i}$ is irrational.

Proof. Because $\lim_{i\to\infty} \frac{t_i}{c_i} = 0$ we can find an index $n > 1$ such that $\frac{t_k}{c_k} < 1$ for all $k \geq n$. We set $T = \sum_{i=1}^{n-1} \frac{t_i}{q_i}$. Because of $q_i = c_1 \cdot c_2 \cdot \dots \cdot c_i \geq 2^i$ we have

$$\begin{aligned}
0 < \alpha = \sum_{i=1}^{\infty} \frac{t_i}{q_i} &= T + \sum_{i=n}^{\infty} \frac{t_i}{q_i} = T + \sum_{i=n}^{\infty} \frac{t_i}{c_1 \cdot c_2 \cdot \dots \cdot c_i} \\
&= T + \sum_{i=n}^{\infty} \frac{1}{c_1 \cdot c_2 \cdot \dots \cdot c_{i-1}} \left(\frac{t_i}{c_i}\right) \\
&< T + \sum_{i=n}^{\infty} \frac{1}{c_1 \cdot c_2 \cdot \dots \cdot c_{i-1}} < T + \left(\frac{1}{2} + \frac{1}{2\cdot 2} + \frac{1}{2\cdot 2\cdot 2} + \cdots\right) \\
&= T + 1
\end{aligned}$$

and we see that α exists since the series converges. Suppose $\alpha = p/q$ could be written as a fraction. Almost like in the proof of Theorem 1, we now choose n so large that $\frac{t_{k+1}}{c_{k+1}} < \frac{1}{2q}$ for all $k \geq n$. We now multiply both sides of the equation $\frac{p}{q} = \sum_{i=1}^{\infty} \frac{t_i}{c_i}$ with $q \cdot q_n$ and obtain

$$\begin{aligned}
& p \cdot q_n - q \cdot \left(\frac{t_1 q_n}{q_1} + \frac{t_2 q_n}{q_2} + \frac{t_3 q_n}{q_3} + \cdots \frac{t_n q_n}{q_n}\right) \\
&= q \sum_{i=n+1}^{\infty} \frac{t_i q_n}{q_i} = q\left(\frac{t_{n+1}}{c_{n+1}} + \frac{t_{n+2}}{c_{n+1}c_{n+2}} + \frac{t_{n+3}}{c_{n+1}c_{n+2}c_{n+3}} + \cdots\right) \\
&= q\left(\left(\frac{t_{n+1}}{c_{n+1}}\right) + \left(\frac{t_{n+2}}{c_{n+1}}\right)\frac{1}{c_{n+2}} + \left(\frac{t_{n+3}}{c_{n+1}}\right)\frac{1}{c_{n+2}c_{n+3}} + \cdots\right) \\
&< \frac{q}{2q}\left(1 + \frac{1}{2} + \frac{1}{2\cdot 2} + \frac{1}{2\cdot 2\cdot 2} + \cdots\right) = 1
\end{aligned}$$

Now we have again obtained the same contradiction as above. The left side is an integer, while the right side lies between zero and one. Again, we must reject our assumption that α can be written as a fraction.

Our conjecture is now that Theorem 2 can be used to show that all roots of natural numbers that are not themselves perfect squares are irrational, not just those that lead to unit fractions.

We now assume that we have found $N = a^2 - b^2$ where a and b are fractions in the general case

$$a = \frac{p}{q}, \quad b = \frac{v}{q}$$

For the further sequence members, the following applies:

$$a' = \frac{2p^2 - v^2}{2pq}, b' = \frac{v^2}{2pq}, a'' = \frac{2(2p^2 - v^2)^2 - v^4}{2pq2(2p^2 - v^2)}, b'' = \frac{v^4}{2pq2(2p^2 - v^2)},$$

$$a''' = \frac{\left(2(2p^2 - v^2)^2 - v^4)^2 - v^8\right)}{2pq2(2p^2 - v^2)2\left(2(2p^2 - v^2)^2 - v^4\right)},$$

$$b''' = \frac{v^8}{2pq2(2p^2 - v^2)2\left(2(2p^2 - v^2)^2 - v^4\right)},$$

and we see that the denominators of the b-sequence have a nested structure, as we need in the above theorem. With the notations from Theorem 2, we have here $t_i = v^{2^i}$ for the numerators. The denominators are somewhat more difficult to define, as we have already seen above. We set $M_1 = p, M_2 = \left(2p^2 - v^2\right), M_3 = 2\left(2p^2 - v^2\right)^2 - v^4$ and $M_{i+1} = 2M_i^2 - v^{2^i}$. Then $c_i = 2M_i$. In order to apply our Theorem 2, we must show that

$$\frac{t_i}{c_i} = \frac{v^{2^i}}{2M_i}$$

is a null sequence. The proof proceeds by induction. It is not obvious that $M_1 = p > 2^{2^1-1}v^{2^1} = 2v^2$. However, we will later show that we can achieve that. So, we will for now assume that $p > 2v^2$. Hence, the statement is correct for $i = 1$.

Assuming that $p > 2v^2$ holds, we can show that $M_i > 2^{2^i-1}v^{2^i}$ holds, which in turn implies that

$$\frac{t_i}{c_i} = \frac{v^{2^i}}{2M_i} < \frac{1}{2^{2^i}} \to 0.$$

Suppose the statement is also valid for a certain index i, i.e. $M_i > 2^{2^i-1}v^{2^i}$, then

$$\begin{aligned} M_{i+1} &= 2M_i^2 - v^{2^i} = 2M_i^2 - M_i + \left(M_i - v^{2^i}\right) \\ &\geq 2M_i^2 - M_i + \left(M_i - 2^{2^i-1}v^{2^i}\right) \geq M_i^2 + M_i(M_i - 1) > 2\left(2^{2^i-1}v^{2^i}\right)^2 \\ &= 2 \cdot 2^{2 \cdot 2^i - 2}v^{2^{i+1}} = 2^{2^{i+1}-1}v^{2^{i+1}}. \end{aligned}$$

Here we have used that $M_i > 2^{2^i-1}v^{2^i}$ implies $M_i - 1 \geq 2^{2^i-1}v^{2^i}$, because these numbers are integers.

Thus, the statement is proved and we only need to show why we can start from $p > 2v^2$. The sizes p and v are, however, only dependent on the initial figure.

So, we must show, that it is always possible to find a starting figure that fulfills $p > 2v^2$.

Very often the condition $p > 2v^2$ is fulfilled from the beginning, e.g. for $\sqrt{19}$ with $a = \frac{35}{8}$ and $b = \frac{3}{8}$. In the following example, however, we see that the condition initially does not hold, but after two iterations it is valid. For this we consider $\sqrt{14}$ and start with $a = \frac{23}{6}$ and $b = \frac{5}{6}$. Here, of course, $a^2 - b^2 = 14$ holds and we have $p = 23$ and $v = 5$. Therefore $p = 23 < 2 \cdot 25 = 50$. After the first iteration we have $a' = \frac{1{,}033}{2 \cdot 6 \cdot 23}$ and $b' = \frac{25}{2 \cdot 6 \cdot 23}$. Here we have $p' = 1{,}033$ and $v' = 25$, and we still have $p' = 1{,}033 < 2 \cdot v'^2 = 2 \cdot 625 = 1{,}250$. After the second step, however, we have $a'' = \frac{2 \cdot 1{,}033^2 - 5^4}{2 \cdot 2 \cdot 6 \cdot 23 \cdot 1{,}033} = \frac{2{,}133{,}553}{2 \cdot 2 \cdot 6 \cdot 23 \cdot 1{,}033}$ and $b'' = \frac{5^4}{2 \cdot 2 \cdot 6 \cdot 23 \cdot 1{,}033} = \frac{625}{2 \cdot 2 \cdot 6 \cdot 23 \cdot 1{,}033}$. Therefore $p'' = 2{,}133{,}553$ and $v'' = 625$, and now we have $p'' = 2{,}133{,}553 > 2 \cdot v''^2 = 2 \cdot 625^2 = 781{,}250$ and the condition is fulfilled.

The second iteration is, of course, already a very good approximation to the root, and it begs the question whether it might not be sensible to start with a very good approximation, i.e., an L-figure that is already very close to a square, if the condition $p > 2v^2$ is to apply from the outset.

We now want to construct such a starting figure. Above, we saw that it can be worthwhile to place N between two consecutive square numbers, i.e., $n^2 < N = n^2 + R < (n+1)^2$. We now refine this idea and form the sequence of growing squares

$$n^2 < \left(n + \frac{1}{2}\right)^2 < \left(n + \frac{3}{4}\right)^2 < \left(n + \frac{7}{8}\right)^2 < \cdots < \left(n + \frac{2^k - 1}{2^k}\right)^2 < \cdots < (n+1)^2$$

and place our N in the correct position in this series. We thus determine an exponent $k \geq 0$, so that

$$\left(n + \frac{2^k - 1}{2^k}\right)^2 < N = n^2 + R < \left(n + \frac{2^{k+1} - 1}{2^{k+1}}\right)^2.$$

As a natural number, N cannot be equal to the left or right boundary, because the lower and upper bounds are proper fractions. In the case of $k = 0$, the left side would be an integer. If N were equal to this left side, then it would already be the square of a integer, which is precisely what is excluded. Therefore, strict inequality must apply on both sides.

Then we have

$$2n\left(\frac{2^k - 1}{2^k}\right) + \left(\frac{2^k - 1}{2^k}\right)^2 < R < 2n\left(\frac{2^{k+1} - 1}{2^{k+1}}\right) + \left(\frac{2^{k+1} - 1}{2^{k+1}}\right)^2$$

and we can construct an L-figure whose area is N. We start from the square $\left(n + \frac{2^k - 1}{2^k}\right)^2$ and attach the remaining area $R - 2n\left(\frac{2^k - 1}{2^k}\right) - \left(\frac{2^k - 1}{2^k}\right)^2 > 0$ again

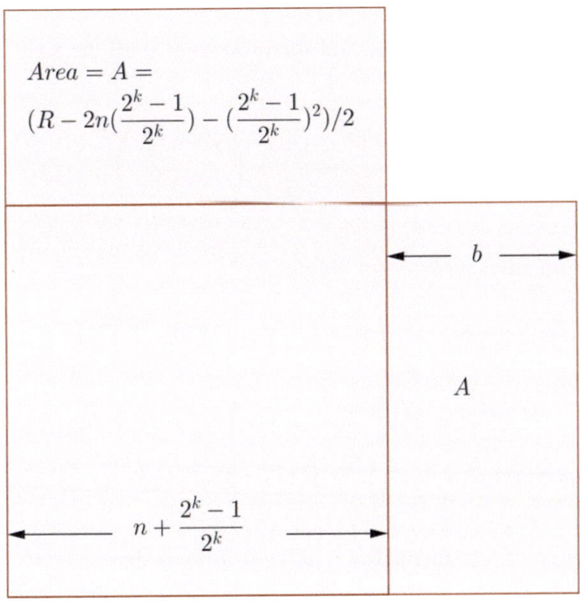

Fig. 2.10 Starting figure

on the outside. This remaining area is distributed over two rectangles of length $\left(n + \frac{2^k-1}{2^k}\right)$. Thus, the width of the rectangles is

$$b = \frac{R - 2n\left(\frac{2^k-1}{2^k}\right) - \left(\frac{2^k-1}{2^k}\right)^2}{2\left(n + \frac{2^k-1}{2^k}\right)},$$

while $a = n + \frac{2^k-1}{2^k} + b$ (see Fig. 2.10).

If we now write this without double fractions, it looks as follows:

$$b = \frac{2^{2k}R - 2^{k+1}n(2^k-1) - \left(2^k-1\right)^2}{2^{k+1}\left(2^k n + 2^k - 1\right)} = \frac{2^{2k}R - 2^{k+1}n(2^k-1) - \left(2^k-1\right)^2}{2^{k+1}\left(2^k(n+1) - 1\right)}$$

and we have found both the numerator $v = 2^{2k}R - 2^{k+1}n(2^k-1) - \left(2^k-1\right)^2$ and the denominator $q = 2^{k+1}\left(2^k(n+1) - 1\right)$ of b.

We can now also calculate the side length a of the large square:

$$a = n + \frac{2^k-1}{2^k} + b = \frac{2^k(n+1) - 1}{2^k} + \frac{v}{2^{k+1}\left(2^k(n+1) - 1\right)}$$
$$= \frac{2\left(2^k(n+1) - 1\right)^2 + v}{2^{k+1}\left(2^k(n+1) - 1\right)}.$$

Thus, the numerator is $p = 2\left(2^k(n+1) - 1\right)^2 + v$.

We want to show that $p > 2v^2$ and therefore start with the right inequality for R, namely

$$R < 2n\left(\frac{2^{k+1} - 1}{2^{k+1}}\right) + \left(\frac{2^{k+1} - 1}{2^{k+1}}\right)^2.$$

We multiply both sides by 2^{2k} and obtain

$$2^{2k}R < 2^k n\left(2^{k+1} - 1\right) + \left(2^{k+1} - 1\right)^2/4$$

$$2^{2k}R < n\left(2^{2k+1} - 2^k\right) + \left(2^{k+1} - 1\right)^2/4$$

$$2^{2k}R - 2^{k+1}n(2^k - 1) - \left(2^k - 1\right)^2 < 2^{k+1}n - 2^k n + \left(2^{k+1} - 1\right)^2/4 - \left(2^k - 1\right)^2.$$

The left hand side of this inequality is v, so it follows that

$$\begin{aligned} v &< 2^k n + \frac{2^{2k+2} - 2^{k+2} + 1 - 2^{2k+2} + 2^{k+3} - 4}{4} \\ &= 2^k n + 2^k - \frac{3}{4} = 2^k(n+1) - \frac{3}{4}, \end{aligned}$$

so $v \leq 2^k(n+1) - 1$, because v, as the numerator of b, is an integer. Thus, we have

$$p = 2\left(2^k(n+1) - 1\right)^2 + v > 2\left(2^k(n+1) - 1\right)^2 \geq 2v^2$$

because $v > 0$. So we have shown that $p > 2v^2$. This means that the sequence of b_i fits into our Theorem 2, and we can conclude that roots of integers that are not themselves squares must be irrational. Thus, we have provided a new proof for the irrationality of roots from natural numbers that are not perfect squares themselves.

2.16 From Integers to Fractions

For fractions $N = \frac{e}{d}$ we can argue as follows: We consider the number $N' = Nd^2$, which is an integer, and calculate the sequence of approximations according to the above pattern. Then $\sqrt{N} = \frac{\sqrt{N'}}{d}$ and we can divide each term in the sum of $\sqrt{N'}$ by d to get an infinite sum for $\sqrt{N}$. Thus, we have reduced the extraction of roots from fractions to the extraction of roots from natural numbers. We also now know that $\sqrt{N'}$ is irrational. The irrationality of $\sqrt{N'}$ is transferred to $\sqrt{N}$. This shows how to extract roots from rational numbers and at the same time proves their irrationality.

Remark 2.6

If we compare, for example, $\sqrt{170} = 13.038404\ldots$ with $\sqrt{1.7} = 1.3038404\ldots$ we see that the answers are composed of the same sequences of digits. This is because $\sqrt{170} = 10\sqrt{1.7}$. It doesn't really matter whether we want to find $\sqrt{170}$ or $\sqrt{1.7}$. If we want to extract the root from a large number N, we can first divide it by 100 or 10,000 to get a smaller number that we can handle more easily. Similarly, a decimal number N below 1 can be multiplied by a power of 100 to bring it to the appropriate order of magnitude for our purposes. N can then, as we have seen above, be a rational number and does not necessarily have to be an integer. We can then assume that we can bring our number N into the interval between 1 and 100 for our considerations. Since N now lies in such a limited interval, it is easy to find the nearest squares $n^2 < N < (n+1)^2$. There are only 9 possibilities, which can be easily found and programed as "if" conditions. Then we also know that $b' < 1/2$ and with $a'^2 - N = b'^2$, it follows that $\left(a' - \sqrt{N}\right)\left(a' + \sqrt{N}\right) = b'^2$. Since $\left(a' - \sqrt{N}\right)$ is the smaller factor in the product, it immediately follows that $\left(a' - \sqrt{N}\right) < b' < 1/2$ and we know that the number of correct decimals doubles in each step, which makes it easier for us to determine the accuracy in advance. So k steps in the algorithm are sufficient to produce 2^k correct decimals.

2.17 The Pell Equation

Bramhagupta (628 AD) studied the Diophantine equation

$$x^2 - Ny^2 = 1,$$

where the solutions of the equation in integers played a special role. Today this equation is named after the mathematician John Pell (1611–1685) [Pell]. He developed a method for generating a third solution from two known solutions of the equation. He based this on the observation of the following identity, which can be confirmed by simple multiplication:

$$\left(x_1^2 - Ny_1^2\right)\left(x_2^2 - Ny_2^2\right) = (x_1x_2 + Ny_1y_2)^2 - N(x_1y_2 + x_2y_1)^2.$$

Let (x_1, y_1) and (x_2, y_2) be two solutions of the above-mentioned Diophantine equation, then

$$u = x_1x_2 + Ny_1y_2 \text{ and } w = x_1y_2 + x_2y_1$$

is also a solution of the equation. If only one solution of the equation is known, a second can still be found by setting $x_1 = x_2$ and $y_1 = y_2$ in the above formula. In particular, we then have

$$u = x_1^2 + Ny_1^2 \text{ and } w = 2x_1y_1.$$

We now want to show how the solutions of the Pell equation are related to the L-figures of the Indian root algorithm. We will see that one iteration step of the Indian algorithm corresponds exactly to the transition from one solution to a second one described by Bramhagupta when combining the solution with itself.

From the Sulbasutra we have $a^2 - N = b^2$, so

$$\left(\frac{a}{b}\right)^2 - N\left(\frac{1}{b}\right)^2 = 1.$$

Thus, $(x_1, y_1) = \left(\frac{a}{b}, \frac{1}{b}\right)$ is a solution pair of the Pell equation. Generally, fractions appear here. If we now use Bramhagupta's method to produce a new solution, we obtain it as

$$u = x_1^2 + Ny_1^2 = \left(\frac{a}{b}\right)^2 + N\left(\frac{1}{b}\right)^2 = \frac{a^2 + N}{b^2} = \frac{2a^2 - b^2}{b^2} \text{ and } w = 2x_1y_1 = \frac{2a}{b^2}.$$

An iteration with the Indian algorithm yields $a' = a - b' = a - \frac{b^2}{2 \cdot a}$ and $b' = \frac{b^2}{2 \cdot a}$, thus

$$\frac{a'}{b'} = \frac{a - \frac{b^2}{2a}}{\frac{b^2}{2a}} = \frac{2a^2 - b^2}{b^2} = u \text{ and } \frac{1}{b'} = \frac{2a}{b^2} = w.$$

Therefore, an iteration step in the Indian algorithm does exactly the same as the application of Bramhagupta's method of a solution pair to itself, and we have demonstrated a connection between the two methods.

We have found that in many examples b is a unit fraction. We have seen this above with $\sqrt{2}$ and $\sqrt{15}$ with our own eyes. There, a and b also have the same denominator. Thus, the corresponding solutions of the Pell equation become integers. If b is a unit fraction, this is also the case for b‘, and all subsequent terms from the b-series. In these cases (b = unit fraction), we have thus solved the Pell equation in integers, and we have also shown how to produce infinitely many solutions.

An example can clarify the situation. We examine the Pell equation

$$x^2 - 2y^2 = 1.$$

After some trial and error, we find the solution $x_1 = 3$ and $y_1 = 2$, because $3^2 - 2 \cdot 2^2 = 1$. Now we apply Bramhagupta's method to find a new solution to our equation. We set $x_2 = x_1^2 + 2 \cdot y_1^2 = 3^2 + 2 \cdot 2^2 = 17$ and $y_2 = 2x_1y_1 = 2 \cdot 3 \cdot 2 = 12$. Then

$$x_2^2 - 2y_2^2 = 17^2 - 2 \cdot 12^2 = 289 - 288 = 1.$$

The values 17 and 12 are familiar to us, because 17/12 is just one of the approximations that occurs in the Indian root method (see also Table 2.1). In the next step, we then get $x_3 = x_2^2 + 2 \cdot y_2^2 = 17^2 + 2 \cdot 12^2 = 577$ and $y_3 = 2x_2y_2 = 2 \cdot 17 \cdot 12 = 408$, also values that we find in Table. 2.1 for the approximations of $\sqrt{2}$ and again

$$x_3^2 - 2y_3^2 = 577^2 - 2 \cdot 408^2 = 332{,}929 - 332{,}928 = 1.$$

With these two new solution pairs, a third solution pair can now be found using Bramhagupta's method, and the search can continue indefinitely.

We will finish by giving one additional example. We examine the Pell equation

$$x^2 - 15y^2 = 1.$$

We see that $x_1 = 4$ and $y_1 = 1$ is a solution, because $4^2 - 15 \cdot 1^2 = 1$. Now we apply Bramhagupta's method to find a new solution to our equation. We set $x_2 = x_1^2 + 15 \cdot y_1^2 = 4^2 + 15 \cdot 1^2 = 31$ and $y_2 = 2x_1y_1 = 2 \cdot 4 \cdot 1 = 8$. Then

$$x_2^2 - 15y_2^2 = 31^2 - 15 \cdot 8^2 = 961 - 960 = 1.$$

In the next step, we then get $x_3 = x_2^2 + 15 \cdot y_2^2 = 31^2 + 15 \cdot 8^2 = 1{,}921$ and $y_3 = 2x_2y_2 = 2 \cdot 31 \cdot 8 = 496$. Again

$$x_3^2 - 15y_3^2 = 1{,}921^2 - 15 \cdot 496^2 = 3{,}690{,}241 - 3{,}690{,}240 = 1$$

and we see that new solutions can always be constructed from known solutions using Bramhagupta's method. At the same time, the relationship with the Indian root method is striking.

2.18 Summary

In this chapter, we have seen that Datta's (1932) and Henderson's (2000) interpretation of an ancient Indian geometric root extraction algorithm can be generalized so that it is applicable to any number. A possible starting figure can always be found. The convergence speed is quadratic, which is an astonishing result for such an old algorithm. It also turns out that the method is equivalent to Heron's method, which is a classic among root algorithms. The Indian method thus gives us a new geometric interpretation of Heron's procedure, which allows us to attach geometric intuition to Heron's method and thus gain a new approach and a new interpretation of the "divide and average" method. It also turns out that the Indian method anticipates the Bramhagupta procedure for Pell equations, which produces infinitely many solutions from one solution.

The Indian method of root extraction is related to a number of other much later discovered procedures, which has led us to particularly appreciate this method and write this chapter.

Bibliography

Datta, B. (1932). *The Science of the Sulbas: A Study in Early Hindu Geometry.* Calcutta, Calcutta University Press, pp. 195–206.

Henderson, D. W. (2000). Square roots in the Sulba Sutra. In C.A. Gorini (ed): *Geometry at Work: Papers in Applied Geometry* , MAA Notes Number 53, pp. 39–45.

Joseph, G. G. (1990). *The Crest of the Peacock.* Penguin Books.

Internet Addresses

[Sulbasutra] https://mathshistory.st-andrews.ac.uk/HistTopics/Indian_sulbasutras/

[Fourier] https://en.wikipedia.org/wiki/Proof_that_e_is_irrational

[Pell] https://en.wikipedia.org/wiki/Pell's_equation

Integration and Differentiation—A Generalization of the Method of Gregorius

3

3.1 Introduction

At this point, a new approach to integral and differential calculus will be introduced, where areas under the graphs of all basic school functions and tangents to these function graphs are determined using elementary geometric means. This is done for power functions—including for $y = 1/x$, exponential functions, the logarithm, and the trigonometric functions sine and cosine. These considerations are made without the fundamental theorem of calculus, but instead they conceptually prepare for it. In the respective sections that describe the individual types of functions, we first deal with area calculation, then tangent calculation. The same geometric tools are used for both.

Historically, there were a variety of methods and approaches for calculating areas under curves, some of which have found their way into textbooks, so that integral calculus sometimes appears unsystematic, before the fundamental theorem of calculus brings "order" back into the matter. However, with the fundamental theorem, we leave a path that has geometric intuition as a guide, because looking at integration as the "reversal" of differentiation, makes it into algebraic and algorithmic calculations. The method presented here works purely geometrically, covers all known school functions, and argues uniformly systematically, without presenting integration as antiderivation. If students are introduced to integration in this way, they learn some geometry and the fundamental theorem is perceived as a geometric fact.

The basic functions of school mathematics, especially power functions with integer exponents, exponential functions, logarithms, and trigonometric functions have certain functional properties. For power functions, we have $(ab)^n = a^n \cdot b^n$, for exponential functions, we have $g^{a+b} = g^a \cdot g^b$, for the logarithm we have $\ln(ab) = \ln(a) + \ln(b)$ and for the trigonometric functions we have the addition theorems. We will see that we can use these functional properties for area and tangent determinations. In addition, we can consider scalings and translations of

© The Author(s), under exclusive license to Springer-Verlag GmbH, DE, part of Springer Nature 2026

C. Kirfel, *Side Paths in the History of Mathematics*, Mathematics Study Resources 21,
https://doi.org/10.1007/978-3-662-72918-2_3

the graphs of the functions to get the same results without the fundamental theorem of calculus.

3.2 The Method of Gregorius of St. Vincent

By the middle of the seventeenth century, Fermat (1601–1665) and Roberville (1602–1675) had already developed formulas for the area under power functions $y = x^k$. The rule

$$\int x^k dx = \frac{x^{k+1}}{k+1} \text{ for } k \neq -1$$

was known. But the hyperbola $f(x) = x^{-1} = 1/x$ was still not understood. Here, Fermat's formula could not be applied, since for $k = -1$ the denominator on the right side becomes zero. The Flemish Jesuit, Gregorius of St. Vincent (Gregorius a San Vincentio 1647), presents in his work *Opus geometricum* of 1647 new ideas for determining areas under hyperbolas, which was the inspiration for the work presented in this chapter (see also Bopp 1907, pp. 241–242 and Volkert 1996). Surprisingly, these ideas are of a general nature, so they can also be transferred to other types of functions. In this chapter, we develop a comprehensive approach for determining the area of all functions that we encounter in school mathematics. We consistently use geometric interpretations and do not let the area calculations "degenerate" into algebraic antidifferentiation using the fundamental theorem of calculus. All this is built on the joint work with R. Kaenders (see also Kaenders 2015, Kaenders and Kirfel 2016, 2017, 2020a, b; Kirfel 2014, 2018). The same approach also allows us to solve the problem of tangent determination with the same tools.

3.3 The Area Under the Hyperba, Gregorius's Approach

We now want to compare the area under the graph of the hyperbola $f(x) = 1/x$ over the interval $[a, b]]$ with that over the interval $[ta, tb]$. We now divide the intervals into smaller subintervals, so that a subpoint t_i in the interval $[a, b]$ corresponds to the subpoint tt_i in the interval $[ta, tb]$. We then compare the associated rectangular subareas (see Fig. 3.1). The rectangular area over the subinterval $\left[t_i, t_{i+1}\right]$ of the interval $[a, b]$, where $a \leq t_i < t_{i+1} \leq b$ has the following area:

$$A = (t_{i+1} - t_i) \cdot \frac{1}{t_i} = \frac{t_{i+1} - t_i}{t_i},$$

while the area of the corresponding rectangle over the subinterval $\left[tt_i, tt_{i+1}\right]$ can be computed in the following way:

$$B = (tt_{i+1} - tt_i) \cdot \frac{1}{tt_i} = \frac{t_{i+1} - t_i}{t_i}.$$

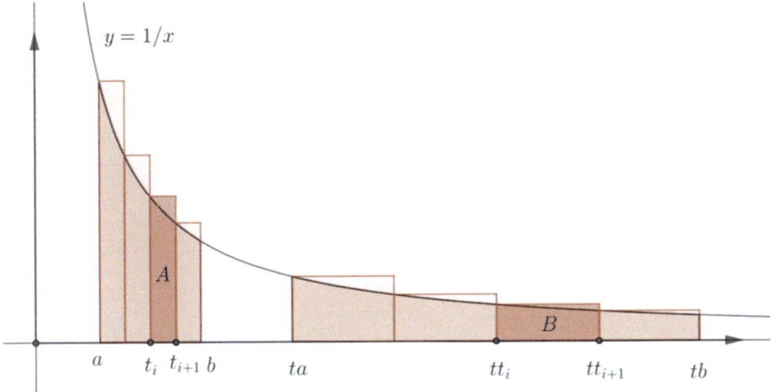

Fig. 3.1 Rectangles of the same area

Surprisingly, the rectangles have the same area. This is a pleasant surprise and makes our work easier.

Then, all the other rectangles over the two intervals must also be pairwise equal in size, and thus the total area must also be the same—even after making the rectangles narrower and taking a limit in order to achieve the area under the hyperbola. Therefore, the area under the hyperbola over the interval $[a, b]$ is equal to the area under the same hyperbola over the interval $[ta, tb]$. This insight is very special and does not occur with other curves. This probably astonished Gregorius of St. Vincent, too.

3.4 Scalings

In this section, we want to deepen the concept of scaling, which we will then make extensive use of in the rest of the chapter. With this, we will be able to gain new insights about areas under curves in the following sections.

First, we imagine a rectangle R in a coordinate system (see Fig. 3.2). The corners here are

$$A = (a, 0), B = (b, 0), C = (b, c) \text{ and } D = (a, c).$$

Next, we perform a scaling in the x-direction with the factor t. This results in the x-coordinate of every point in the plane, including the corners of the quadrilateral, being multiplied by t. The y-values of the points remain unchanged. The corners of the scaled quadrilateral R' are then

$$A' = (at, 0), B' = (bt, 0), C' = (bt, c) \text{ and } D' = (at, c).$$

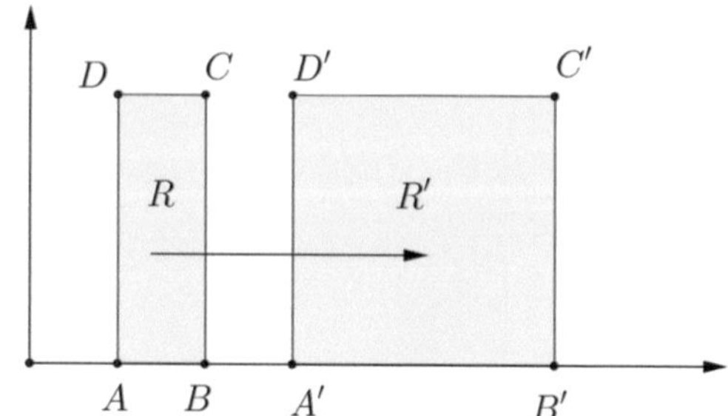

Fig. 3.2 Scaling in x-direction

We can imagine that the rectangle is made of a rubber sheet, and that we have pulled on the rubber sheet to bring it into the new shape. It is also immediately apparent that the area of the scaled rectangle is t times as large as the original. Similarly, we can also work with scalings in the direction of the y-axis (see Fig. 3.3). Here too, the area grows to t times the original area, if the scaling factor is t.

What about areas that have curved boundaries, such as the hyperbola? Next, we consider the scaling of a hyperbola with scaling factor $t = 3$. We imagine that the curved area under the lower curve ($f(x) = 1/x$) from $x = 1$ to $x = 2$ is composed of many rectangles.

Each of these rectangles is transformed into a new scaled rectangle under the upper curve ($y = t/x = 3/x$) during the scaling. The area of the individual rectangles increases by a factor of t. Then, the total area under the curve (here from 1 to 2) also increases by a factor of t (here it triples). The interval boundaries also shift from 1 to 3 and from 2 to 6 (see Fig. 3.4).

In a scaling with the factor t, the area thus grows to t times its original size. This naturally also applies to scalings in the y-direction.

How can we now understand the transition from the lower curve to the upper curve? Each point $\left(x, \frac{1}{x}\right)$ of the lower curve is mapped to the point $\left(tx, \frac{1}{x}\right)$ of the

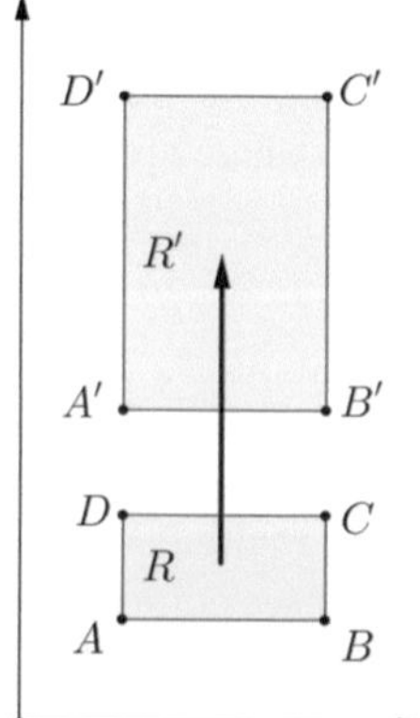

Fig. 3.3 Scaling in y-direction

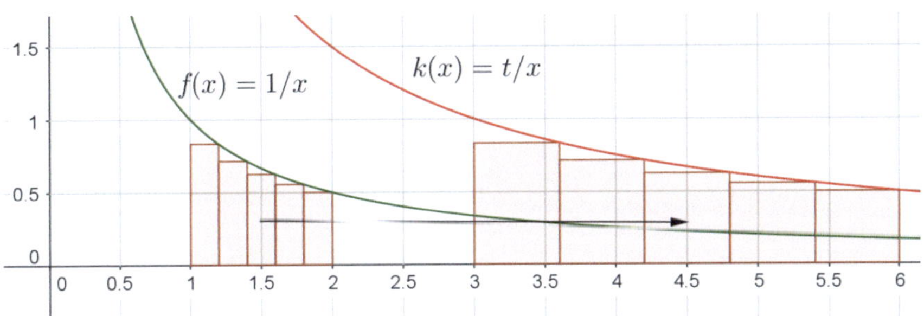

Fig. 3.4 Scaling of a hyperbola

upper curve. For the points of the upper curve $(\overline{x}, \overline{y})$, the following applies:

$$\overline{x} = tx \quad \text{and} \quad \overline{y} = y = \frac{1}{x} = \frac{1}{\left(\frac{\overline{x}}{t}\right)} = \frac{t}{\overline{x}}.$$

The formula of a scaled (upper) curve $\overline{y} = t/\overline{x}$ is obtained by replacing x with x/t in the original formula.

If a scaling is performed with the scaling factor μ in the direction of the y-axis, we must accordingly replace y with $\frac{y}{\mu}$. We get $\overline{x} = x$ and $\overline{y} = \mu y$. If the original curve represents a hyperbola $y = 1/x$, then the scaled curve has the formula $\overline{y} = \mu y = \frac{\mu}{x} = \frac{\mu}{\overline{x}}$. For a scaling in the direction of the y-axis, we only need to multiply the right side of the formula for y by the scaling factor μ. This gives us an overview of how we can describe the scalings purely in terms of formulas. For the rest of this chapter, we will for simplicity dispense with the distinction between y and $\overline{y}$ and also x and $\overline{x}$.

3.5 The Area Under the Hyperbola with Scalings

We now want to show the result that the areas under a hyperbola over the intervals $[a, b]$ and $[ta, tb]$ are equal from the second perspective, namely through considerations with scalings. The observation of what happens during a scaling does not have to be limited to rectangles, but can, as we have just seen, directly refer to the entire bounded area under the graph of the function $y = \frac{1}{x}$.

We consider the area A under the function $f(x) = 1/x$ over the interval $[a, b]$. By scaling along the x-axis with the factor t, we obtain the function $k(x) = (t/x)$ and the area $\overline{A}$ under the new function k over the scaled interval $[ta, tb]$ is $\overline{A} = t \cdot A$. As we have seen above, a scaling with a factor t generally also changes the area by the factor t (see Fig. 3.5).

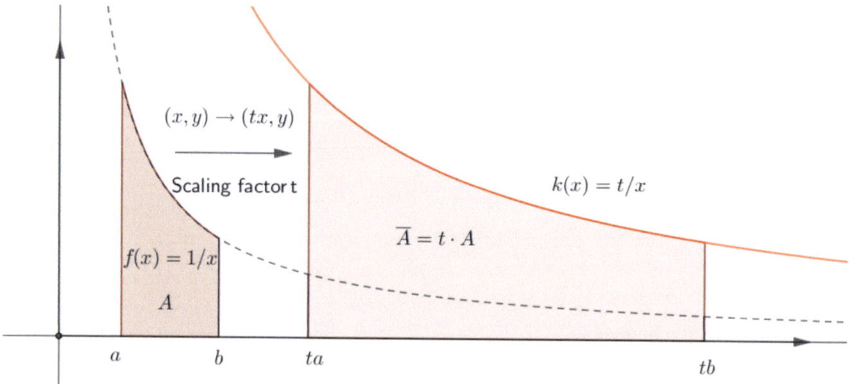

Fig. 3.5 Scaling of a hyperbola in x-direction

We now perform a new scaling, this time in the direction of the y-axis. The scaling factor is this time μ. The new function h fulfills the following condition:

$$h(x) = \mu k(x) = \mu(t/x) = \mu t/x = (\mu t)f(x).$$

In Fig. 3.6 we see the three functions. We now claim that by choosing the second scaling factor μ cleverly, namely $\mu = 1/t$, the new function $h(x)$ can "be pressed down" to coincide with the original function $f(x)$.

This second scaling in the y direction, which is carried out with the factor $\mu = 1/t$, could also be called a "compression". If $t > 1$ then $\mu = 1/t < 1$, and the area decreases in the transition.

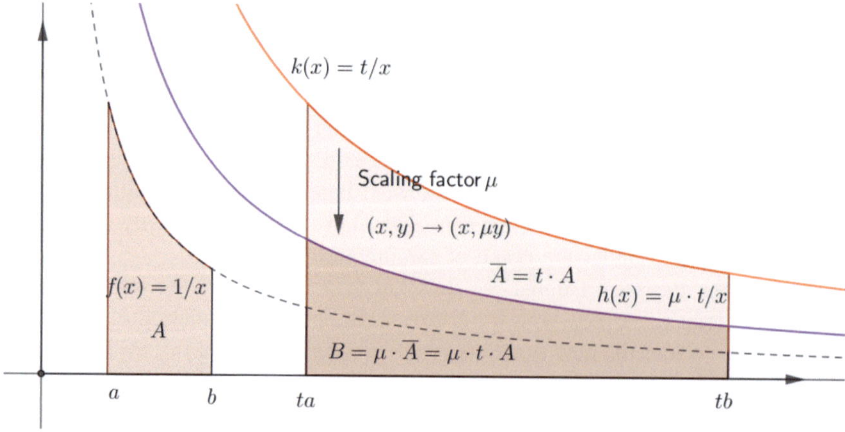

Fig. 3.6 Scaling of a hyperbola in y-direction

For the area B over the extended interval $[ta, tb]$ under the new function h, which is now equal to the original function f, we have

$$B = t^{-1} \cdot \overline{A} = A,$$

and the areas under the hyperbola over the intervals $[a, b]$ and $[ta, tb]$ are equal as above. This is the same result as Gregorius obtained. Next, we want to show where the connection to the logarithmic function comes from. Let's set, for example, $T = d^m$ for a base d and an integer exponent m. Then the following applies

$$m = \log_d T.$$

With this, we can write

$$\begin{aligned} L(T) &= \int_1^T \frac{dx}{x} = \int_1^{d^m} \frac{dx}{x} = \int_1^d \frac{dx}{x} + \int_d^{d^2} \frac{dx}{x} + \cdots + \int_{d^{m-1}}^{d^m=T} \frac{dx}{x} \\ &= m \int_1^d \frac{dx}{x} = C \cdot \log_d T, \end{aligned}$$

and we clearly see that L is a logarithmic function (see Fig. 3.7). Here we put $C = \int_1^d \frac{dx}{x}$.

Remark 3.1
Gregorius himself probably did not fully see the connection of his discovery with the logarithm, but his contemporary Sarasa may have understood this connection. For details, see Burn (2001), Burn (1999), and Edwards (1979). The method has gained new popularity through Toeplitz (1949).

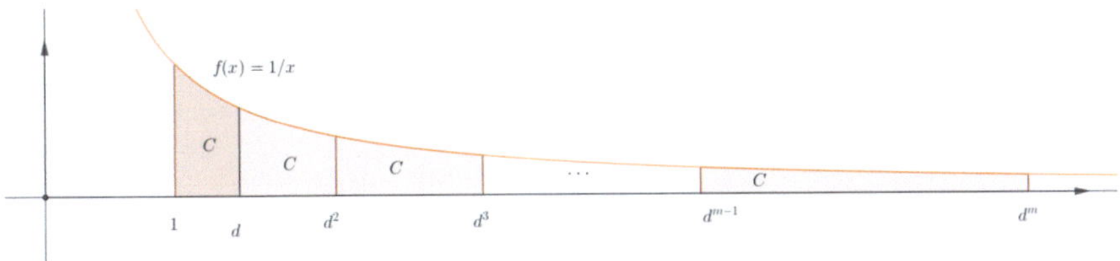

Fig. 3.7 The logarithm as the integral of the hyperbola

3.6 Consequences for the Area Under the Hyperbola

The property that the areas under the hyperbola over the intervals $[a, b]$ and $[ta, tb]$ are equal can now be used to introduce the exponential function, as is common in analysis. This can also be done in school. However, the theoretical arguments must be explained through intuitive considerations. First, for two positive numbers a, b:

$$\int_1^{ab} \frac{dx}{x} = L(ab) = \int_1^{a} \frac{dx}{x} + \int_a^{ab} \frac{dx}{x} = \int_1^{a} \frac{dx}{x} + \int_1^{b} \frac{dx}{x} = L(a) + L(b),$$

which is just the logarithmic property of the function $L(x)$. We also see that the area function $L(x)$ grows strictly monotonically. In Fig. 3.8 we see that the area under the hyperbola grows indefinitely. We have

$$\frac{1}{2} + \left(\frac{1}{3} + \frac{1}{4}\right) + \left(\frac{1}{5} + \frac{1}{6} + \frac{1}{7} + \frac{1}{8}\right) + \cdots > \frac{1}{2} + \frac{1}{2} + \frac{1}{2} + \cdots.$$

We can repeatedly add new blocks of fractions, whose sum is more than ½. But we can do this as often as we like, which means that the area under the hyperbola grows to infinity. By symmetry, this also applies to the area under the hyperbola to the left of $x = 1$.

Since $f(x) = 1/x$ is continuous and positive, the area function must also be continuous and strictly monotonically increasing. As usual, the intermediate value theorem then guarantees the existence of an inverse function exp and it can be shown that it has the property $\exp(x + y) = \exp(x) \cdot \exp(y)$ for all $x, y \in \mathbb{R}$.

For the logarithmic function $L(x)$ the following applies for any positive number g and integer m

$$L(g^m) = mL(g),$$

and for fractions $\frac{p}{q}$ we have

$$qL\left(g^{\frac{p}{q}}\right) = L(g^p) = pL(g), \text{ so } L\left(g^{\frac{p}{q}}\right) = \frac{p}{q}L(g).$$

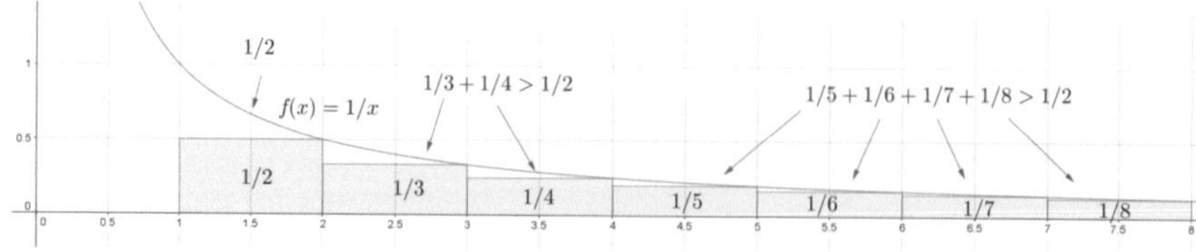

Fig. 3.8 The area under the hyperbola is unbounded

Since the area function $L(x)$ is continuous, it follows that

$$L(g^x) = xL(g) \quad \text{for all} \quad x \in \mathbb{R}.$$

Now let's define the function $h(x) = \exp(L(g)x)$. For integer m we get

$$h(m) = \exp(L(g)m) = (\exp(L(g)))^m = g^m$$

and for fractions $\frac{p}{q}$ we have

$$h(p/q) = \exp(L(g)p/q) = (\exp(L(g)))^{p/q} = g^{p/q}$$

and the functions g^x and $h(x) = \exp(L(g)x)$ agree for all rational values and thus for all $x \in \mathbb{R}$. Here again $g^{x+y} = g^x \cdot g^y$ holds for all $x, y \in \mathbb{R}$.

This means that the function $h(x) = g^x = \exp(L(g)x))$ is precisely the inverse function of the function L if $L(g) = 1$. We now want to find this special base g. We will now consider the area difference between g and tg for t close to 1. This equals

$$\int_1^{gt} \frac{dx}{x} - \int_1^{g} \frac{dx}{x} = L(gt) - L(g) = L(t),$$

which can be seen as a narrow rectangle in Fig. 3.9. Due to the monotonicity of the hyperbola, we can easily estimate this difference using rectangular boxes.

Then it follows that

$$\frac{gt - g}{gt} \leq L(t) \leq \frac{gt - g}{g} \quad \text{or} \quad \frac{1}{t} \leq \frac{L(t)}{t - 1} \leq 1.$$

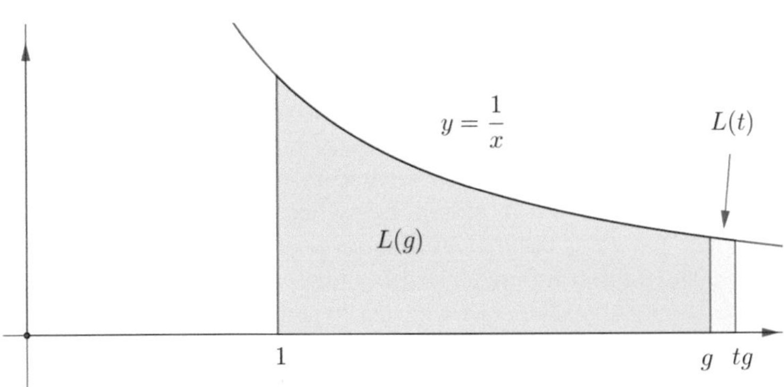

Fig. 3.9 Narrow strip under the hyperbola

We now set $t = 1 + \frac{1}{n}$ and obtain

$$\frac{1}{1+\frac{1}{n}} \le nL\left(1+\frac{1}{n}\right) \le 1 \quad \text{or} \quad g^{\frac{1}{1+1/n}} \le \left(1+\frac{1}{n}\right)^n \le g.$$

This means that the required base g must be the Euler number

$$e = \lim_{n\to\infty}\left(1+\frac{1}{n}\right)^n$$

, and the function $L(x) = \ln(x)$ is the natural logarithm.

From Fig. 3.9 we could see that

$$\frac{1}{t} \le \frac{L(t)}{t-1} = \frac{L(t)-L(1)}{t-1} \le 1.$$

Thus, we have also found the derivative of the natural logarithm function at the point $(1, 0)$. It has the value 1. Due to the mirror symmetry for inverse functions, the slope of the tangent to the Euler function $f(x) = e^x$ at the mirror point $(0, 1)$ also has the value 1. We will use both of these facts in the course of the chapter.

3.7 Tangents to the Hyperbola

Leon van den Broek (1994) describes how to determine the tangent to a hyperbola. In this part of the chapter we will extend this method, which also relies on simple geometric transformations (scalings), to determine tangents to all the classical functions of school (power functions, exponential functions, logarithm and the trigonometric functions, sine and cosine). This opens up a new approach to calculus, based exclusively on simple geometric transformations, translations and scalings. These new perspectives can possibly be incorporated in school but certainly in teacher education.

As when we computed the integral, it is again the hyperbola that opens up this access. The counterpart to the hyperbola integration of Gregorius is the determination of the tangent to the hyperbola of Leon van den Broek (1994).

Van den Broek (1994) presents a method in his article where he calculates the slope of the tangent to a hyperbola by scaling the hyperbola (along with the tangent) twice. We were inspired by this presentation. We start from a known tangent slope and transfer it through simple geometric transformations to any other point on the same curve. We start with the hyperbola $f(x) = \frac{1}{x}$. For reasons of symmetry, the tangent $y = -x+2$ to the hyperbola $f(x) = \frac{1}{x}$ at the point $(1, 1)$ (see Fig. 3.10, left) is known. Now we scale the entire plane along with the hyperbola and tangent from the y-axis along the x-axis by the factor t, i.e., we use the scaling $(x, y) \mapsto (tx, y)$. This gives us a new hyperbola $k(x) = \frac{t}{x}$ and a new tangent at the point $(t, 1)$. To compute the slope of this tangent, we observe that the slope triangle associated with the first tangent, with leg lengths 1 and 1, is transformed by the scaling into the new slope triangle with leg lengths t and 1. The leg that ran parallel to the x-axis is scaled by a factor of t, while the leg that ran parallel to the y-axis remains unchanged. Therefore, the new slope is $-\frac{1}{t}$ (see Fig. 3.10, middle below). Now we perform a second scaling of the plane along with the new hyperbola and the new tangent, this time in the direction of the y-axis. The scaling factor is $\frac{1}{t}$, i.e., $(x, y) \mapsto \left(x, \frac{1}{t}y\right)$. The function scaled for the second time is then

$$h(x) = \frac{1}{t} \cdot k(x) = \frac{1}{t} \cdot \frac{t}{x} = \frac{1}{x} = f(x).$$

i.e., we have returned to the original function. The tangent, scaled for the second time, then has the slope $-\frac{1}{t} \cdot \frac{1}{t} = -\frac{1}{t^2}$ and touches the function graph at the point $\left(t, \frac{1}{t}\right)$ (see Fig. 3.10, right). In the second scaling, the horizontal leg of the slope triangle was retained, while the vertical leg of the slope triangle was scaled by the factor $\frac{1}{t}$. Thus, we have found the slope of the tangent at any arbitrary point x, namely $f'(x) = -\frac{1}{x^2}$.

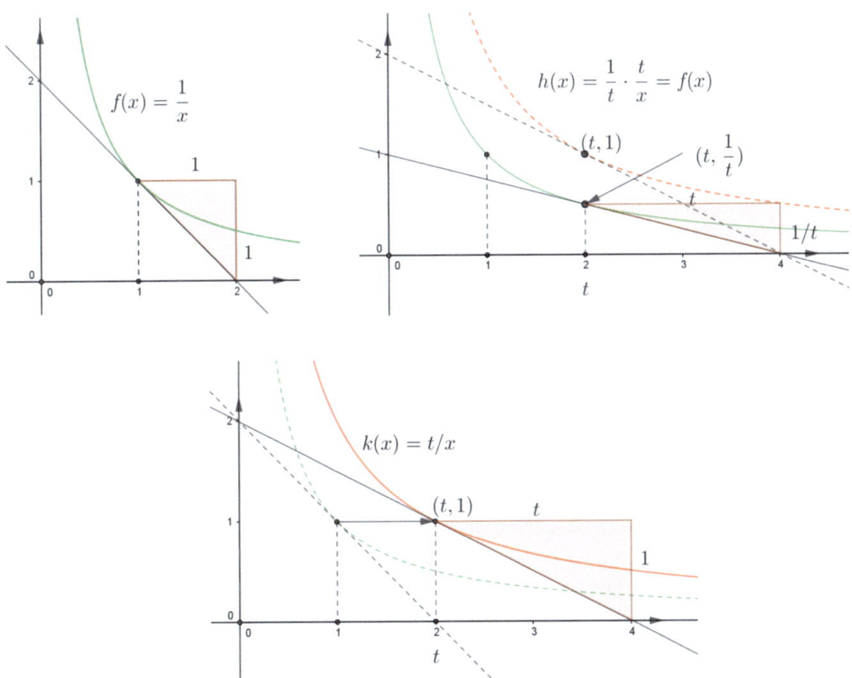

Fig. 3.10 Double scaling of the hyperbola and its tangent at the point of symmetry

3.8 Generalizing to Other Functions

Gregorius's (1647) idea was to consider the area under the hyperbola over a given interval $[a, b]$, and then compare it with the area under the same hyperbola, but over another interval $[ta, tb]$, where the x-coordinates of the endpoints had been multiplied by the same number t. This comparison provides important insights into the formulas for the sought-after areas. In this comparison, we can basically take two different paths, which lead to the same insights. One, where we consider functional properties and one where we consider scalings and other geometric transformations. Therefore, our presentation will be dual-track at this point, so that we present the different perspectives in parallel and thus easily comparable in a synoptic overview. We will also follow the same dual-track approach when calculating the derivatives.

It turs out that the method can be used for areas under curves of all the functions that typically occur in school (power functions, exponential functions, logarithms, and trigonometric functions). The area under the curve over a certain interval is always compared with the area over another interval. We use the term "Gregorius's Method", although Gregorius himself did not design this method in this way. But that is precisely the aim of this book to show where classical methods have potential that could not yet be exploited at the time of discovery, but where we can make further progress today—with the help of modern notation and modern tools, such as coordinate system and concept of function. However, Gregorius's ideas about areas under hyperbolas have encouraged us to generalize his approach, so the naming after him is justified.

Where we want to determine the derivatives algebraically using the functional properties, we will resort to limit considerations, which sometimes are based on a "multiplicative" approach instead of an "additive approach" to the infinitesimal increase ($f(rx_0) - f(x_0)$ instead of $f(x_0 + h) - f(x_0)$).

In the geometric considerations, we again use scalings and translations. It turns out that it is quite easy to study the impact of scaling a function on the derivative by studying the associated slope triangles. As a result, we only need to determine the derivative at a single point directly via the local rate of change of the function and then transfer it to all other points using such transformations.

3.9 Areas Under Power Functions Through Calculus with Their Functional Property

Next, we look at the power function $y = x^n$. The area of a narrow strip over $\left[t_i, t_{i+1}\right]$ in the interval $[a, b]$ is compared to the corresponding area of a strip over $\left[tt_i, tt_{i+1}\right]$ in the interval $[ta, tb]$. The first area is

$$A = (t_{i+1} - t_i)t_{i+1}^n,$$

while the second is

$$B = (tt_{i+1} - tt_i)(tt_{i+1})^n = t^{n+1}A$$

is, just a multiple of A (see Fig. 3.11). Since this is true for such pairs of strips, the same now also applies to the entire area over $[ta, tb]$ compared to the area over $[a, b]$. Therefore, we can write

$$\int_{ta}^{tb} x^n dx = t^{n+1} \int_{a}^{b} x^n dx.$$

We use this formula to show that:

$$\int_{0}^{t} x^n dx - \int_{t\cdot 0}^{t\cdot 1} x^n dx = t^{n+1} \int_{0}^{1} x^n dx = C_P \cdot t^{n+1}.$$

Here $C_P = \int_0^1 x^n dx$ is a constant, namely the area under the curve between 0 and 1. Thus, we have shown that the area function for a power function is again a power function, but this time of a higher degree. The only property of the power function that we used here is the following $(tx)^n = t^n \cdot x^n$. Notice that we must assume $n \neq -1$, in order to start integration from 0. We will compute C_P in the next section.

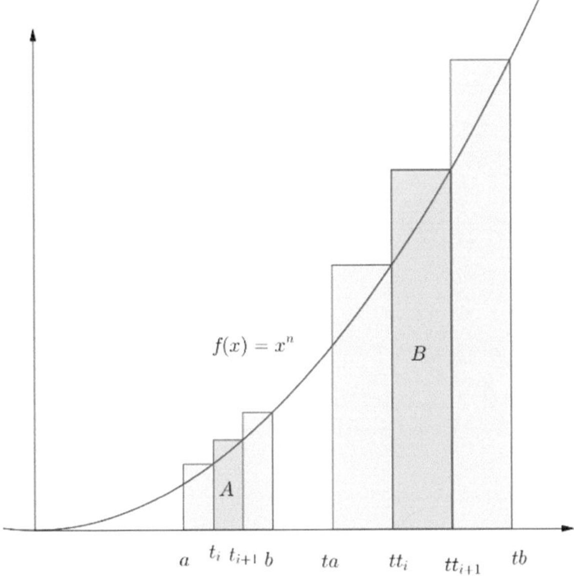

Fig. 3.11 The area under the power function

3.10 Areas Under Power Functions with Double Scaling

We now want to show the result for the power functions from the second perspective, namely through considerations with scalings (see Fig. 3.12). We consider the area A under the function $f(x) = x^n$ over the interval $[a, b]$. By scalings along the x-axis by a factor of t, we obtain the function $k(x) = (x/t)^n$ and the area $\overline{A}$ under the new function k over the stretched interval $[ta, tb]$ is $\overline{A} = t \cdot A$.

We now perform a new scaling, this time in the direction of the y-axis. The scaling factor this time is μ. The resulting function h fulfills the following condition:

$$h(x) = \mu k(x) = \mu(x/t)^n = (\mu/t^n)x^n = (\mu/t^n)f(x).$$

If we now choose the second scaling factor μ wisely, namely $\mu = t^n$ then the new function $h(x)$ coincides with the original function $f(x)$ (see Fig. 3.13).

For the area B over the scaled interval $[ta, tb]$ under the new function h, which is now equal to the original function f, the following applies:

$$B = t^n \cdot \overline{A} = t^{n+1} \cdot A, \text{ so as above } \int_0^t x^n dx = C_p \cdot t^{n+1}.$$

If we set here $n = -1$ we get $A = B$ and the formula again confirms the special result for the hyperbola, namely, that the areas under the hyperbola over the intervals $[a, b]$ and $[ta, tb]$ are equal. However, when $n = -1$, we cannot start integration at 0.

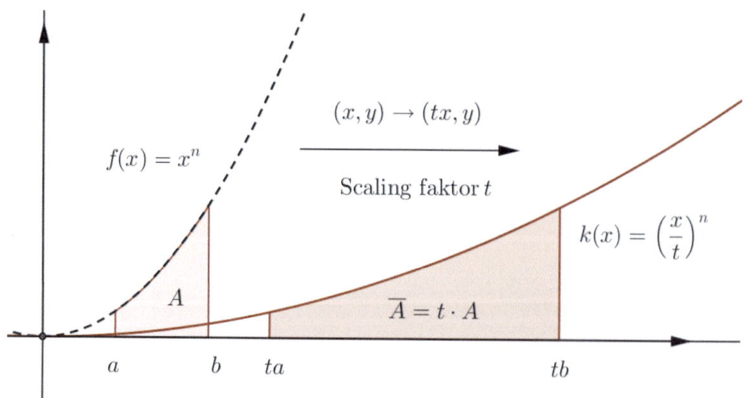

Fig. 3.12 Scaling of the power function

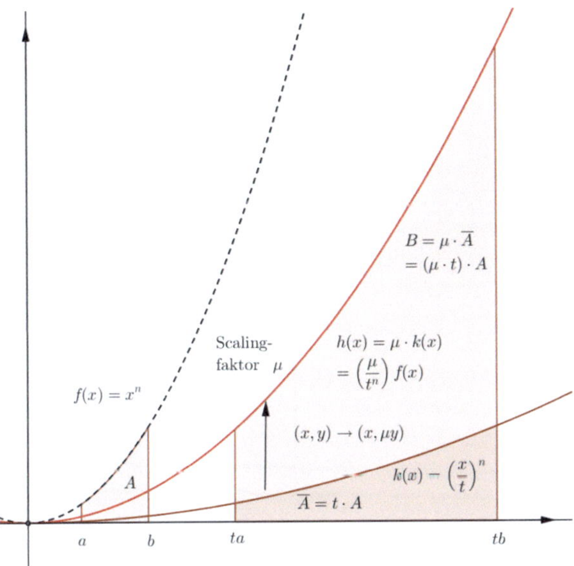

Fig. 3.13 Second scaling of the power function

The integral formula is the same as in the previous section and we will now calculate the constant C_P. To do this, we calculate the areas over the intervals $[0, 1]$ and $[0, t]$ for $1 < t$, with t close to 1. The difference between these areas,

$$\int_0^t x^n dx - \int_0^1 x^n dx = C_P t^{n+1} - C_P,$$

which we can recognize as a narrow strip in Fig. 3.14, we now estimate from above and below with rectangular strips. Here we use the monotonicity of the power function. The result is

$$\begin{aligned}(t-1) \cdot 1 &\le C_P t^{n+1} - C_P \le (t-1) \cdot t^n \\ 1 &\le \frac{t^{n+1}-1}{t-1} C_P \le t^n \\ 1 &\le \left(1 + t + t^2 + \ldots + t^n\right) C_P \le t^n.\end{aligned}$$

Since this holds for all values of $1 < t$, we get $C_P = \frac{1}{n+1}$ for $n \in \mathbb{N}$.

Remark 3.2
The method presented here (except for the calculation of the limit C_P) is valid for all real values of $n \neq -1$, not just for natural exponents. In the section on the derivative

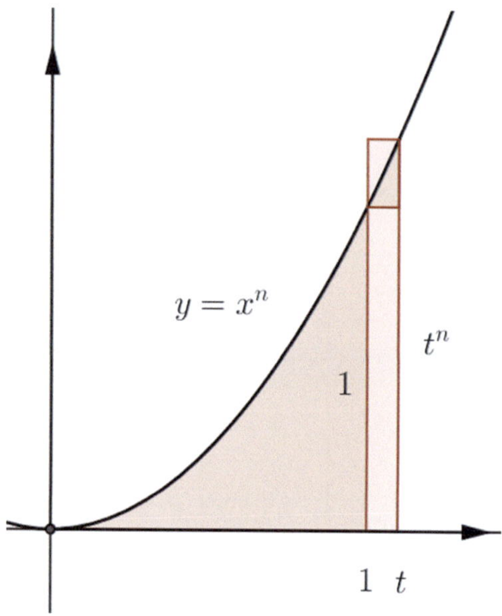

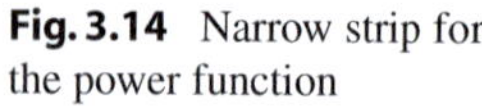
Fig. 3.14 Narrow strip for the power function

of power functions below, we will also show that the limit calculation is not only valid for natural exponents n but is generally valid.

3.11 The Derivative of Power Functions Through Calculus with Their Functional Properties

In many representations of the difference quotient in textbooks, we see an "additive approach". Involuntarily and often automatically, textbook authors start as follows: $f'(x) = \lim_{h\to 0} \frac{f(x+h)-f(x)}{h}$, as if we could only imagine the infinitesimal increase as an additive increase. This often shapes the entire further course of the calculations. However, if we consider the special functional properties of a function, a different approach often presents itself.

If we change the definition only slightly, namely to $f'(x) = \lim_{z\to x} \frac{f(z)-f(x)}{z-x}$, i.e., an understanding that a point z adjacent to x moves towards x in some way, this opens up, for example, a view for a multiplicative understanding; with the approach $z = r \cdot x$ we get:

$$f'(x) = \lim_{r\to 1} \frac{f(r\cdot x) - f(x)}{r\cdot x - x}.$$

With this seemingly insignificant shift in emphasis, in the case of power functions, the above result can be stated without needing to invoke the binomial theorem, which would be the case if an additive interpretation were used. For $f(x) = x^n$, we have

$$f'(x) = \lim_{r\to 1} \frac{f(rx) - f(x)}{x(r-1)} = \lim_{r\to 1} \frac{(rx)^n - x^n}{x(r-1)} = x^{n-1} \lim_{r\to 1} \frac{r^n - 1}{r-1} = x^{n-1} \cdot f'(1).$$

Here, we lack the derivative at a single point, namely at $x = 1$. This means: if we know the derivative at a single point $(1, f(1))$, we also know it at every other point. The case of the hyperbola for the exponent $n = -1$ was already discussed at the beginning. There, due to symmetry, the slope of the tangent at the point (1,1) was known. This is not the case with other power functions. Here, we need to calculate a specific limit value:

$$f'(1) = \lim_{r\to 1} \frac{f(r) - f(1)}{r-1} = \lim_{r\to 1} \frac{r^n - 1}{r-1} = \lim_{r\to 1} \left(r^{n-1} + r^{n-2} + \cdots + 1\right) = n.$$

We had already found this limit value in Sect. 3.10 on the area calculation for power functions. From this, the well-known result on the derivative of power functions follows:

$$f'(x) = x^{n-1} f'(1) = nx^{n-1}.$$

This argument initially only applies to natural numbers n. For $n = 0$, the argument is trivially valid and for negative integer values $n = -\nu$ we have

$$\begin{aligned}\lim_{r\to 1} \frac{r^n - 1}{r-1} &= \lim_{r\to 1} \frac{r^{-\nu} - 1}{r-1} = \lim_{r\to 1} \frac{1}{r^\nu} \cdot \frac{1 - r^\nu}{r-1} \\ &= -\lim_{r\to 1} \frac{1}{r^\nu} \cdot \left(r^{\nu-1} + r^{\nu-2} + \cdots + 1\right) = -\nu = n.\end{aligned}$$

Thus, the limit calculation applies to all integers. But if $n = p/q$ is a fraction, we can argue similarly. Here, we set $w = r^{1/q}$ and use that $w \to 1$, when $r \to 1$. Then we get

$$\begin{aligned}\lim_{r\to 1} \frac{r^n - 1}{r-1} &= \lim_{r\to 1} \frac{r^{p/q} - 1}{r-1} = \lim_{w\to 1} \frac{w^p - 1}{w^q - 1} = \lim_{w\to 1} \frac{w^p - 1}{w - 1} \cdot \frac{w-1}{w^q - 1} \\ &= \lim_{w\to 1} \frac{w^{p-1} + w^{p-2} + \cdots + 1}{w^{q-1} + w^{q-2} + \cdots + 1} = \frac{p}{q} = n\end{aligned}$$

Thus, the calculation is also correct for fractions $n = p/q$. For irrational exponents n, we can use the relationship $x^n = \exp(n\ln(x))$ to show that $(x^n)' = (\exp(n\ln(x)))' = \exp(n\ln(x)) \cdot n \cdot \frac{1}{x} = x^n \cdot n \cdot \frac{1}{x} = nx^{n-1}$, provided we know the chain rule and the differentiation of the exponential function (see Sects. 3.19 and 3.20).

3.12 The Derivation of Power Functions Using Scalings

Again, we start with the function $f(x) = x^n$ for any n, which does not necessarily have to be an integer, and at the point $(1, f(1))$ we consider the tangent with the slope $f'(1)$. We now perform a scaling of the entire plane along with the function along the x-axis with the factor t (Fig. 3.15, left and center).

In this process, both the function graph and the slope triangle are correspondingly scaled. The horizontal side of the triangle is scaled with factor t, while the vertical side does not change. The slope of the new function $k(x)$ at the point t is therefore $k'(t) = \frac{f'(1)}{t}$. Our new function $k(x)$ has now the form $k(x) = \left(\frac{x}{t}\right)^n$ due to the scaling.

Now we scale again—this time in the direction of the y-axis (Fig. 3.15, center, right). Now let the scaling factor be $\mu > 0$. The new function h, that arises, fulfills the following condition:

$$h(x) = \mu \cdot k(x) = \mu \cdot \left(\frac{x}{t}\right)^n = \left(\frac{\mu}{t^n}\right)x^n = \left(\frac{\mu}{t^n}\right)f(x).$$

If we now choose the second scaling factor μ wisely, namely $\mu = t^n$, then the new function $h(x)$ coincides with the original function $f(x)$.

This allows us to calculate the slope of the tangent at h or f at the point $(t, h(t)) = (t, f(t))$. We then have

$$f'(t) = h'(t) = \mu \cdot k'(t) = \left(\frac{\mu}{t}\right)f'(1) = t^{n-1}f'(1),$$

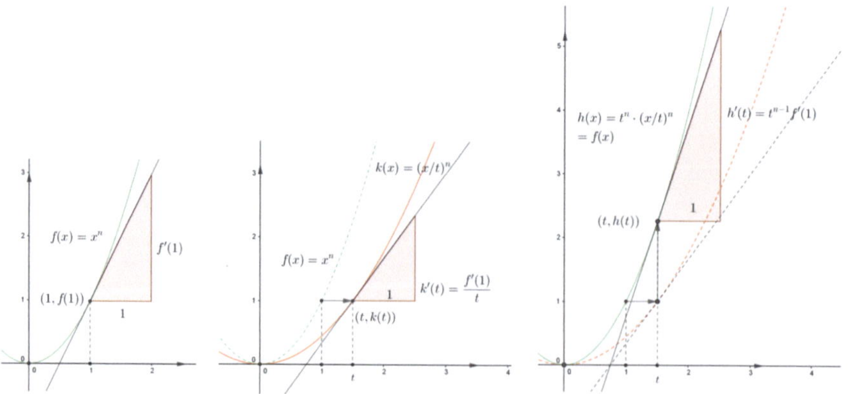

Fig. 3.15 Scaling of the power function first along the x-axis and then along the y-axis together with the tangents and the slope triangles

where we have chosen $\mu = t^n$. Above, we have already calculated this missing derivative value. It follows that $f'(1) = n$. From this, we also obtain the well-known result

$$f'(x) = f'(1 \cdot x) = x^{n-1} f'(1) = nx^{n-1}.$$

3.13 The Area Under the Logarithmic Function by Calculus with Its Functional Property

A very similar argument for finding the area as with the power function can also be used for the logarithmic function. Here we have $A = (t_{i+1} - t_i)\ln(t_{i+1})$ for the area of the strip in the first interval, while the area of the strip in the corresponding second interval can be calculated as

$$B = (tt_{i+1} - tt_i)\ln(tt_{i+1}) = t(t_{i+1} - t_i)(\ln(t_{i+1}) + \ln(t)) = tA + t\ln(t)(t_{i+1} - t_i)$$

(see Fig. 3.16). The new area is now more than a multiple of the old one. We get an additional term $t\ln(t)(t_{i+1} - t_i)$. Now, t is a constant and does not change in the small intervals $\left[t_i, t_{i+1}\right]$, which together make up the interval $[a, b]$. Therefore, it is easy to specify the total value of these additional terms, where all the small intervals are added together. In total, we get: $t\ln(t)(b - a)$. Therefore, we can write:

$$\int_{ta}^{tb} \ln(x)dx = t\int_{a}^{b} \ln(x)dx + t\ln(t)(b - a).$$

We can use this formula to derive the integral of the logarithm function, but before that we will show how we can obtain this formula using scalings and translations.

3.14 The Area Under the Logarithmic Function with Scaling

We would now like to show the result for the logarithmic function from the second perspective, namely through considerations with scaling. We consider the area A under the function $f(x) = \ln(x)$ over the interval $[a, b]$. By scaling along the x-axis with the factor t we get the function $k(x) = \ln\left(\frac{x}{t}\right) = \ln(x) - \ln(t)$ and the area $\overline{A}$ under the new function k over the scaled interval $[ta, tb]$ is $\overline{A} = t \cdot A$.

Now the vertical distance between the functions $f(x) = \ln(x)$ and $k(x) = \ln(x) - \ln(t)$ is constant since t is constant and we can easily calculate the missing area between $f(x)$ and $k(x)$ over the interval $[ta, tb]$ (see Fig. 3.17).

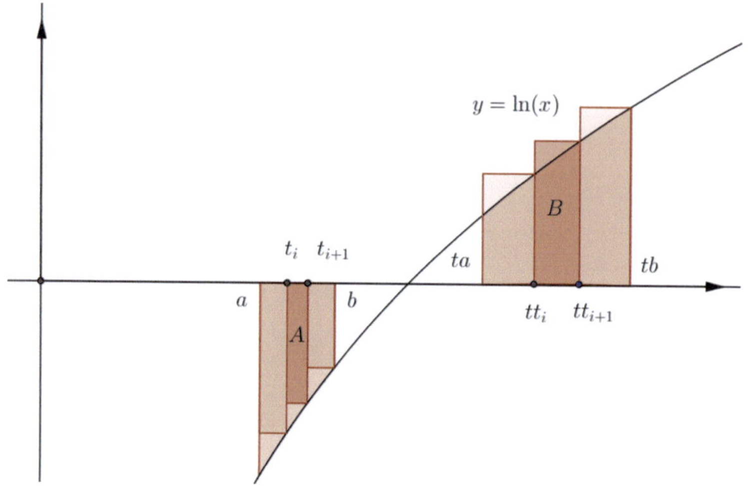

Fig. 3.16 The area under the logarithm function

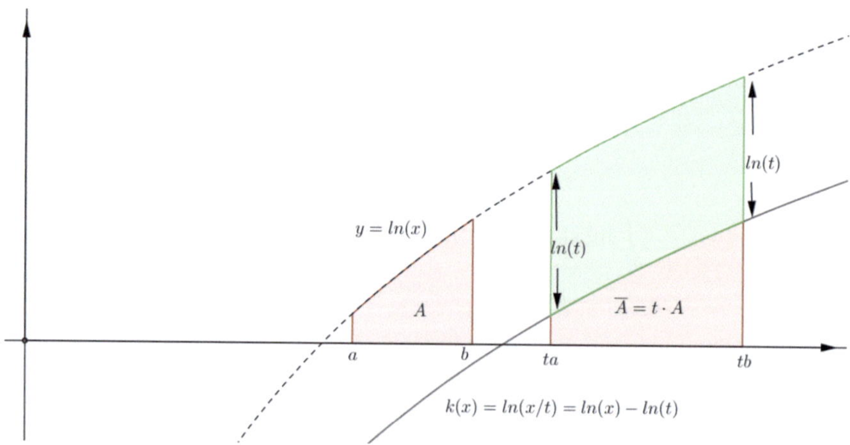

Fig. 3.17 Computation of the area under the logarithm function with scalings and translations

Therefore, we obtain as above

$$B = t \cdot A + \ln(t) \cdot (bt - at) = t \cdot A + t\ln(t)(b - a).$$

We now use this relationship to show the following:

$$\int_0^t \ln(x)dx = \int_{t\cdot 0}^{t\cdot 1} \ln(x)dx = t \int_0^1 \ln(x)dx + t\ln(t)(1 - 0) = C_L \cdot t + t\ln(t),$$

where $C_L = \int_0^1 \ln(x)dx$ is a constant, this time the area "under" the curve between 0 and 1. At first, it is not clear whether this constant C_L even exists, as it describes an "infinitely long" area, i.e., an improper integral. Later we will see that the mentioned area is finite and that $C_L = -1$ holds. With this, we obtain the well-known formula for the logarithmic integral

$$\int_0^t \ln(x)dx = -t + t\ln(t).$$

3.15 The Derivative of the Logarithmic Function by Calculus with Its Functional Property

Let's now examine the slope of the tangent at the logarithmic function. Again, we would like to start from a "multiplicative" understanding of the infinitesimal increase due to the functional properties $\ln(a \cdot b) = \ln(a) + \ln(b)$. For $f(x) = \ln(x)$ we have

$$\begin{aligned} f'(x) &= \lim_{r \to 1} \frac{f(rx) - f(x)}{x(r-1)} = \lim_{r \to 1} \frac{\ln(rx) - \ln(x)}{x(r-1)} \\ &= \lim_{r \to 1} \frac{\ln(r) + \ln(x) - \ln(x)}{x(r-1)} \\ &= \frac{1}{x} \cdot \lim_{r \to 1} \frac{\ln(r) - 0}{r - 1} = \frac{f'(1)}{x}. \end{aligned}$$

We are only missing the derivative $f'(1)$ at a single point, namely $x = 1$. Above, in Sect. 3.6 (Fig. 3.9) on the area calculation under the logarithmic function, we have seen that $\ln'(1) = 1$. With this, we know the slope of the tangent at the point $(1, 0)$. Thus, we have

$$f'(x) = \ln'(x) = \frac{1}{x}.$$

3.16 The Derivative of the Logarithmic Function with Scaling

We now again perform a scaling along the x-axis with the factor t and obtain the function $k(x) = \ln\left(\frac{x}{t}\right) = \ln(x) - \ln(t)$. The corresponding slope of the tangent at the point $(t, k(t))$ is therefore

$$k'(t) = \frac{f'(1)}{t} = \frac{1}{t}.$$

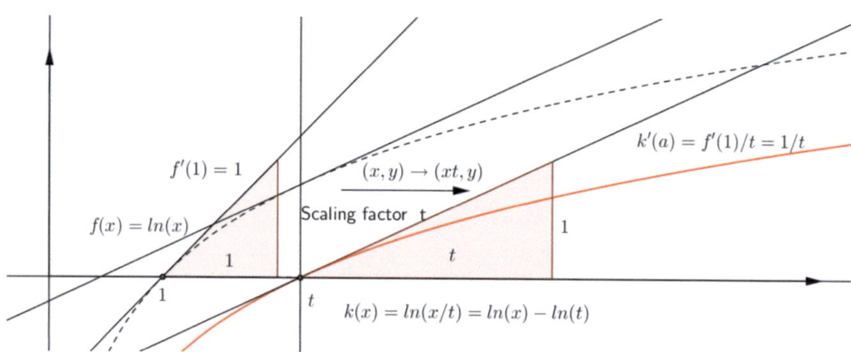

Fig. 3.18 Scaling of the plane with the graph of the logarithm along the x-axis

Here too, we can imagine the slope triangle corresponding to the first tangent (see Fig. 3.18). During the scaling, the horizontal side of the triangle is scaled, while the vertical side does not change. The new slope is therefore

$$k'(t) = \frac{f'(1)}{t} = \frac{1}{t}.$$

However, since the graph of our function $k(x) = \ln\left(\frac{x}{t}\right) = \ln(x) - \ln(t)$ compared to the graph of the original function $f(x) = \ln(x)$ is only translated downwards, where the derivative—i.e., the slope of the tangent—remains unchanged, we have:

$$f'(t) = k'(t) = \frac{f'(1)}{t} = \frac{1}{t}$$

3.17 Areas Under Exponential Functions Through Calculus with Their Functional Properties

The method can also be applied to an exponential function $f(x) = g^x$. Here, the idea is to compare the area over an interval with the area over a translated (not a scaled) interval of the same width (see Fig. 3.19).

We compare the area over $[a, b]$ with that over $[a + t, b + t]$. Within these intervals, we now consider the strips over $\left[t_i, t_{i+1}\right]$ and $\left[t_i + t, t_{i+1} + t\right]$. Here

$$A = (t_{i+1} - t_i)g^{t_{i+1}}$$

describes the area of the strip in the first interval, while

$$B = ((t_{i+1} + t) - (t_i + t))g^{t_{i+1}+t} = g^t(t_{i+1} - t_i)g^{t_{i+1}} = g^t A,$$

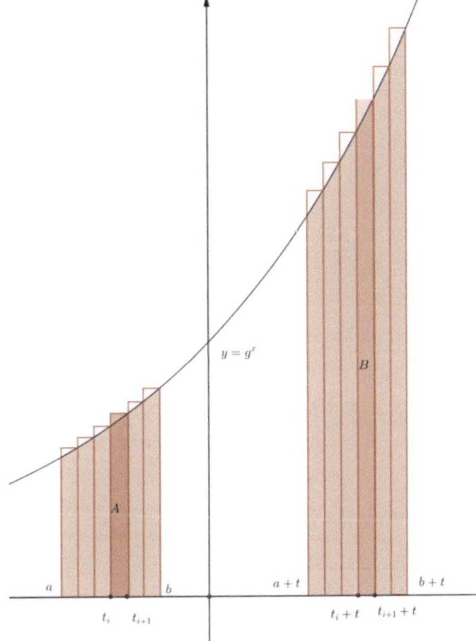

Fig. 3.19 Computation of the area under the exponential functions

describes the area of the strip in the second interval (both darkly colored). Again, the new area is merely a multiple of the old one, which must then also apply to the total area over the intervals $[a+t, b+t]$ and $[a, b]$, because the same factor appears in each case. Then we can write:

$$\int_{t+a}^{t+b} g^x dx = g^t \int_a^b g^x dx.$$

This gives us:

$$\int_{-\infty}^{t} g^x dx = \int_{-\infty+t}^{0+t} g^x dx = g^t \int_{-\infty}^{0} g^x dx = C_E g^t.$$

Here, $C_E = \int_{-\infty}^{0} g^x dx$ is again a constant, this time the area under the curve from $-\infty$ to 0. At first, it is not clear whether this area is finite at all.

We consider Fig. 3.20. Here we see that the mentioned area F under the graph of the function f is smaller than the sum of the areas of the rectangles. So

$$0 < F < 1 \cdot 1 + 1 \cdot \frac{1}{g} + 1 \cdot \frac{1}{g^2} + 1 \cdot \frac{1}{g^3} + \cdots = \frac{1}{1 - \frac{1}{g}} = \frac{g}{g-1},$$

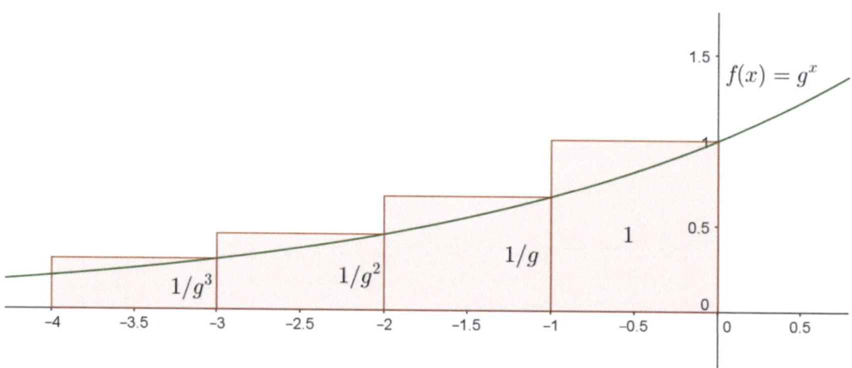

Fig. 3.20 The finiteness of the area under the exponential function to the left of the y-axis

for $g > 1$ and thus the area is finite and the quantity C_E is well-defined. We have used the fact that an increasing sequence of real numbers, which is bounded, has a limit.

Thus, we have shown that the integral of an exponential function is again an exponential function with the same base. The only property of the exponential function that we used here is $g^{a+b} = g^a \cdot g^b$.

3.18 Areas Under Exponential Functions with Translations and Scalings

We now want to show the result for the exponential functions from the second perspective, namely using scalings and translations. We consider the area A under the function $f(x) = g^x$ over the interval $[a, b]$. This time we translate the interval and the function graph by the amount t. The new interval then looks like this $[a + t, b + t]$ and the new function k then has the form $k(x) = g^{x-t}$. The area $\overline{A}$ is unchanged under the new function k over the translated interval $[a + t, b + t]$ (see Fig. 3.21).

We now perform a scaling in the direction of the y-axis. The scaling factor is μ. The resulting new function h satisfies the following condition:

$$h(x) = \mu k(x) = \mu g^{x-t} = \left(\mu/g^t\right)g^x = \left(\mu/g^t\right)f(x).$$

If we now choose the scaling factor μ cleverly, namely $\mu = g^t$ then the new function $h(x)$ coincides with the original function $f(x)$.

For the area B over the translated interval $[a + t, b + t]$ under the new function h, which is now equal to the original function f, the following applies, just as we have shown above:

$$B = g^t \cdot \overline{A} = g^t \cdot A$$

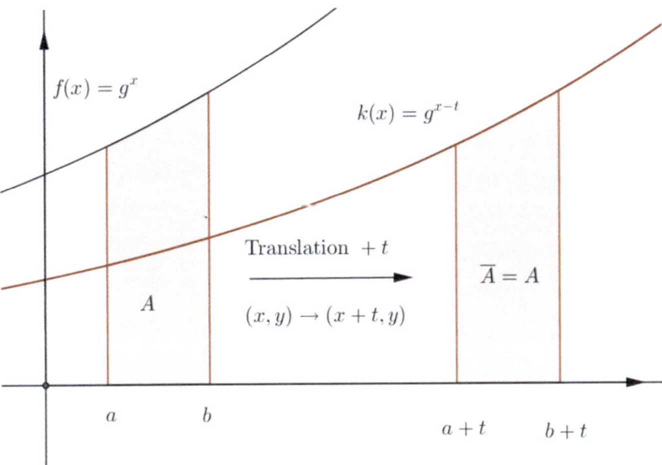

Fig. 3.21 Areas under the exponential function

and again, the area first translated and then scaled is just a multiple of the original area (see Fig. 3.22) and we get as above

$$\int_{t+a}^{t+b} g^x dx = g^t \int_a^b g^x dx.$$

Thus

$$\int_{-\infty}^{t} g^x dx = \int_{-\infty+t}^{0+t} g^x dx = g^t \int_{-\infty}^{0} g^x dx = C_E g^t,$$

where we again put $\int_{-\infty}^{0} g^x dx = C_E$. Above we showed that the constant C_E, although it is defined as an improper integral, exists, i.e. is finite.

We now want to try to establish a connection between the constant C_E and the base g for the exponential function. We are particularly interested in specifying the base g that gives us the simplest and most practical value for the constant, namely $C_E = 1$. To do this, we can proceed similarly to the power function and the logarithm and consider the area difference

$$\int_{-\infty}^{t} g^x dx - \int_{-\infty}^{0} g^x dx = C_E g^t - C_E,$$

where $t > 0$ is close to zero, which can be seen in Fig. 3.23 as a narrow strip. We estimate up and below by rectangles using the monotony of the exponential

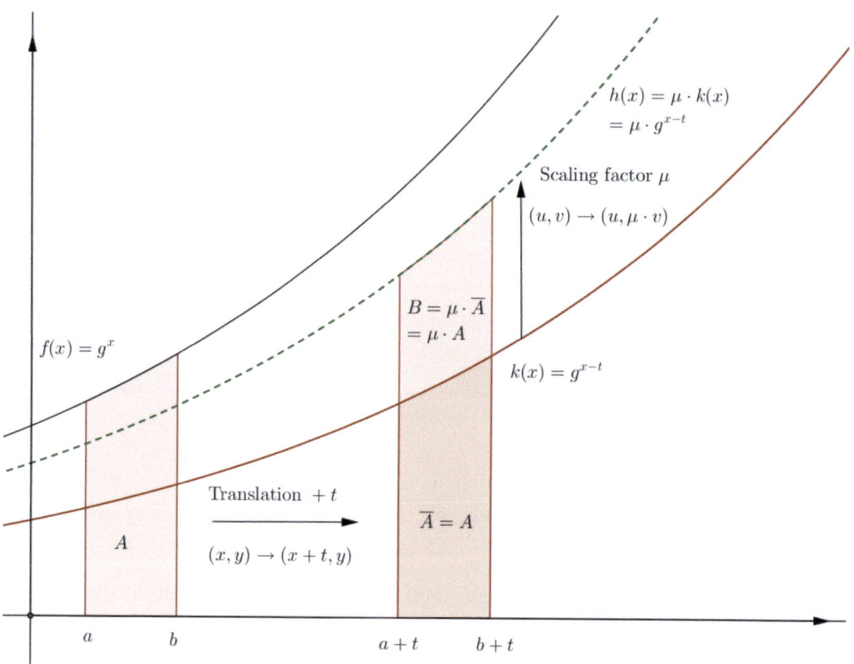

Fig. 3.22 Translation and scaling of the area under the exponential function

function. This results in

$$t \cdot 1 \leq C_E g^t - C_E \leq t \cdot g^t.$$

If we assume $C_E = 1$ and want to determine g, we have

$$t \cdot 1 \leq g^t - 1 \leq t \cdot g^t.$$

We now consider the two inequalities separately. On the left side we get $1+t \leq g^t$. On the right side we get $g^t(1-t) \leq 1$ or $g^t \leq \frac{1}{1-t}$. We now set $t = \frac{1}{n}$ and get

$$1 + \frac{1}{n} \leq g^{\frac{1}{n}} \leq \frac{1}{1 - 1/n}.$$

By exponentiating we get

$$\left(1 + \frac{1}{n}\right)^n \leq g \leq \left(1 + \frac{1}{n-1}\right)^n$$

and we see that the desired base is again provided by the Euler number e.

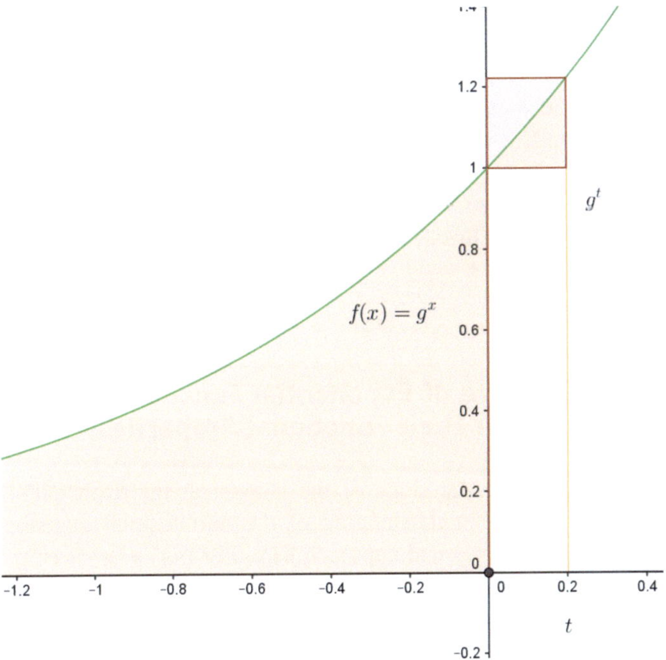

Fig. 3.23 Narrow strip under the exponential function

Above we saw that $t \cdot 1 \le g^t - 1 \le t \cdot g^t$ in the case that $C_E = 1$ or $g = e$. But that means

$$t \cdot 1 \le e^t - 1 \le t \cdot e^t \text{ or } 1 \le \frac{e^t - 1}{t} \le e^t.$$

Thus, we have $\lim_{t \to 0} \frac{e^t - 1}{t} = 1$ and the function $y = e^x$ has slope 1 at $x = 0$. For a general base g, we then have

$$t \cdot 1 \le C_E g^t - C_E \le t \cdot g^t$$

$$\frac{t}{g^t - 1} \le C_E \le \frac{t}{g^t - 1} g^t$$

Thus, here is

$$\begin{aligned} C_E &= \lim_{t \to 0} \frac{t}{g^t - 1} = \lim_{t \to 0} \frac{t}{e^{\ln(g)t} - 1} \\ &= \frac{1}{\ln(g)} \lim_{t \to 0} \frac{\ln(g)t}{e^{\ln(g)t} - 1} = \frac{1}{\ln(g)} \lim_{w \to 0} \frac{w}{e^w - 1} = \frac{1}{\ln(g)} \end{aligned}$$

and we have also determined the constant C_E in the general case.

We have already shown above that the function $L(x)$, which describes the area under the hyperbola, is the inverse function of the exponential function to the base e. We can finally calculate the constant $C_L = \int_0^1 \ln(t)dt$ that was missing above in Sect. 3.14. This constant C_L measures the area between the graph of the natural logarithm function $y = L(x)$ and the y-axis and the x-axis. Therefore, it is equal to the area under Euler's exponential function from $-\infty$ to zero, because this exponential function is the inverse function of the natural logarithm. Therefore, we have $C_L = -C_E = -1$. The negative sign comes from the fact that the mentioned area lies below the x-axis.

3.19 The Derivative of Exponential Functions Through Calculus with Their Functional Properties

We now want to examine the slope of the tangent at the exponential function. At this point, the additive understanding of the infinitesimal increase fits best. Therefore, we follow the classical approach here. For $f(x) = g^x$ it is

$$f'(x) = \lim_{h\to 0} \frac{f(x+h) - f(x)}{h} = \lim_{h\to 0} \frac{g^{x+h} - g^x}{h} = g^x \cdot \lim_{h\to 0} \frac{g^h - g^0}{h} = f(x) \cdot f'(0).$$

Here we only lack the derivative at a single point, namely $x = 0$.

Above we have seen that Euler's exponential function, $f(x) = e^x$ is the inverse function of the natural logarithm—which itself was defined as the integral of the $y = 1/x$. There we also showed that $f'(0) = 1$. Therefore, $f'(x) = f(x) = e^x$ holds for Euler's exponential function.

This allows us to find the derivative of general exponential functions $p(x) = g^x$ with any base g, since

$$\begin{aligned} p'(x) &= \lim_{h\to 0} \frac{g^{x+h} - g^x}{h} = g^x \lim_{h\to 0} \frac{g^h - 1}{h} = g^x \lim_{h\to 0} \frac{e^{h\ln(g)} - 1}{h} \\ &= g^x \ln(g) \lim_{h\to 0} \frac{e^{h\ln(g)} - 1}{h \cdot \ln(g)} = g^x \ln(g) \lim_{w\to 0} \frac{e - 1}{w} = g^x \ln(g). \end{aligned}$$

3.20 The Derivative of Exponential Functions with Translations and Scalings

We now want to examine the slope of the tangent at an exponential function using scalings and translations. We start with the function $f(x) = g^x$. At the point $(0, 1)$ we consider the tangent with the slope $f'(0)$. This time we do not perform a scaling, but a translation along the x-axis, where the translation vector has the length t. We then get the function $k(x) = g^{x-t}$ and the corresponding tangent slope at the point $(t, k(t))$ remains unchanged due to the translation, so $k'(t) = f'(0)$. The associated slope triangle does not change during the translation (see Fig. 3.24, middle).

We now perform a scaling in the direction of the y-axis (Fig. 3.24, right). The scaling factor is μ. The new function h fulfills the following condition:

$$h(x) = \mu \cdot k(x) = \mu \cdot g^{x-t} = \left(\frac{\mu}{g^t}\right) g^x = \left(\frac{\mu}{g^t}\right) f(x).$$

If we now choose the stretch factor μ wisely, namely $\mu = g^t$, then the new function $h(x)$ coincides with the original function $f(x)$. We now calculate the slope of the tangent at h or f at the point $(t, h(t)) = (t, f(t))$, where we have chosen $\mu = g^t$. We then have

$$f'(t) = h'(t) = \mu k'(t) = \mu f'(0) = g^t f'(0).$$

That is: If we know the derivative at a single point (0, 1), we also know it at every other point. From this we get the well-known result as above

$$f'(x) = f'(x+0) = g^x f'(0).$$

The computation of the factor $f'(0)$ is found in Sect. 3.19.

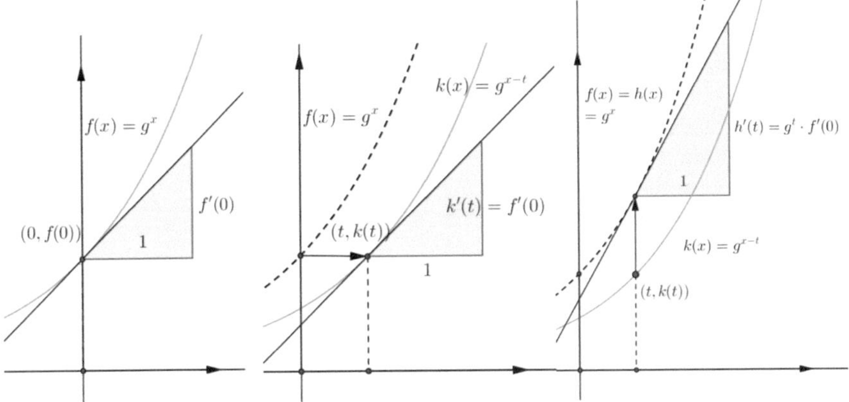

Fig. 3.24 Translation and scaling of the graph of an exponential function

3.21 Areas Under Trigonometric Functions Through Calculus with Their Functional Properties

The method also works for calculating the area under the sine function. We compare the areas over the intervals $[0, \pi]$ and $[t, \pi + t]$. Here

$$A = (t_{i+1} - t_i)\sin(t_i)$$

describes the area of a strip in the first interval $[0, \pi]$ while the area of a strip in the second (translated) interval $[t, \pi + t]$ looks as follows:

$$\begin{aligned} B &= ((t_{i+1} + t) - (t_i + t))\sin(t_i + t) \\ &= (t_{i+1} - t_i)(\sin(t_i)\cos(t) + \cos(t_i)\sin(t)) \\ &= \cos(t)A + (t_{i+1} - t_i)\cos(t_i)\sin(t), \end{aligned}$$

see Fig. 3.25.

Here we have

$$\begin{aligned} \int_t^{\pi+t} \sin(x)dx &= \cos(t)\int_0^{\pi} \sin(x)dx + \sin(t)\int_0^{\pi} \cos(x)dx \\ &= \cos(t)\int_0^{\pi} \sin(x)dx = C_S\cos(t), \end{aligned}$$

because $\int_0^{\pi} \cos(x)dx = 0$. Here we set $\int_0^{\pi} \sin(x)dx = C_S$. From Fig. 3.26 we see that $\int_t^{t+\pi} \sin(x)dx = C_S - 2\int_0^t \sin(x)dx$.

In total we get

$$\int_0^t \sin(x)dx = \left(\frac{C_S}{2}\right)(1 - \cos(t)).$$

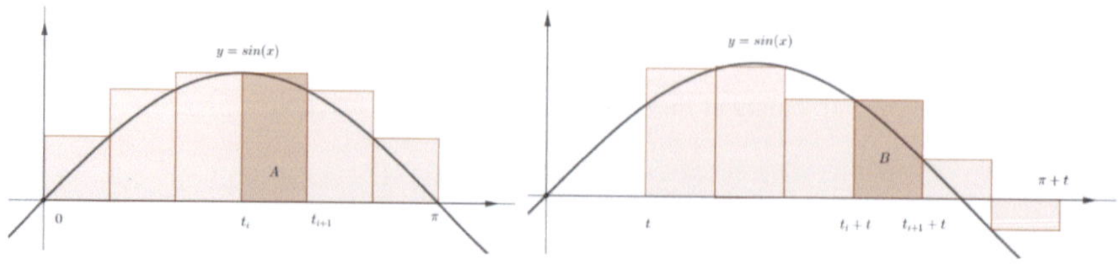

Fig. 3.25 The area under the sine function

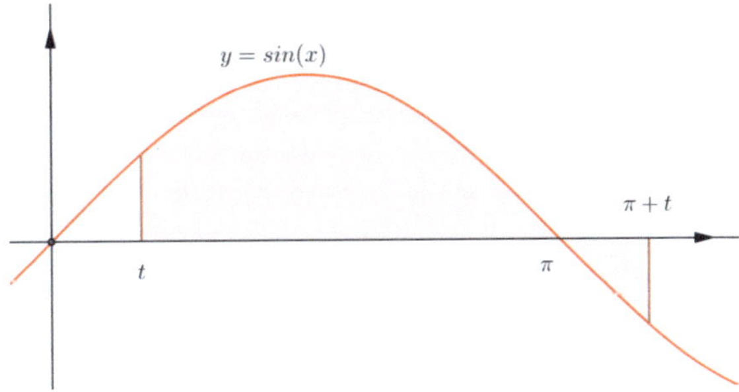

Fig. 3.26 A property of the sine function

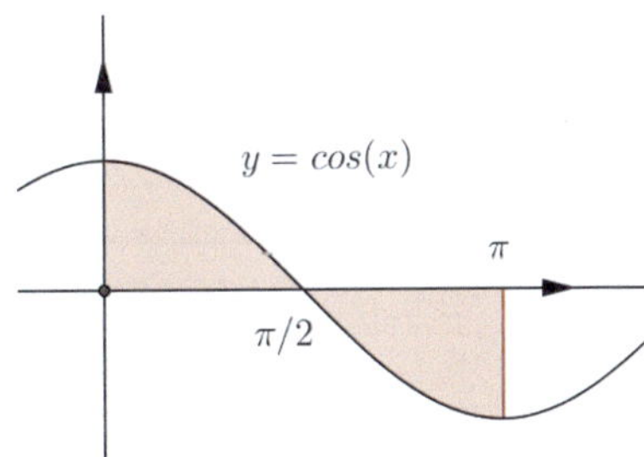

Fig. 3.27 The areas cancel each other

In this argument, we only used the following property of the sine function:

$\sin(a+b) = \sin(a)\cos(b) + \cos(a)\sin(b)$, in particular $\sin(a+\pi) = -\sin(a)$. We also used that: $\cos(a+b) = \cos(a)\cos(b) - \sin(a)\sin(b)$ in particular $\cos\left(\frac{\pi}{2}+a\right) = -\sin(a) = -\cos\left(\frac{\pi}{2}-a\right)$, which gives us $\int_0^\pi \cos(x)dx = 0$ (see Fig. 3.27).

3.22 Areas Under Trigonometric Functions with Translations and Scalings

We now want to show the result for the trigonometric functions from the second perspective, namely by considering scalings and translations (see Fig. 3.28). We consider the area A under the function $f(x) = \sin(x)$ over the interval $[t, t+\pi]$. This time we translate the interval by $-t$ and get the interval $[0, \pi]$. The new function k then has the form

$$k(x) = \sin(x+t) = \sin(x)\cos(t) + \cos(x)\sin(t)$$

and for the area $\overline{A}$ under the new function k over the translated interval $[0, \pi]$, which is equal to the initial area A, we have

$$A = \int_t^{t+\pi} \sin(x)dx = \overline{A} = \int_0^\pi \sin(x+t)dx$$

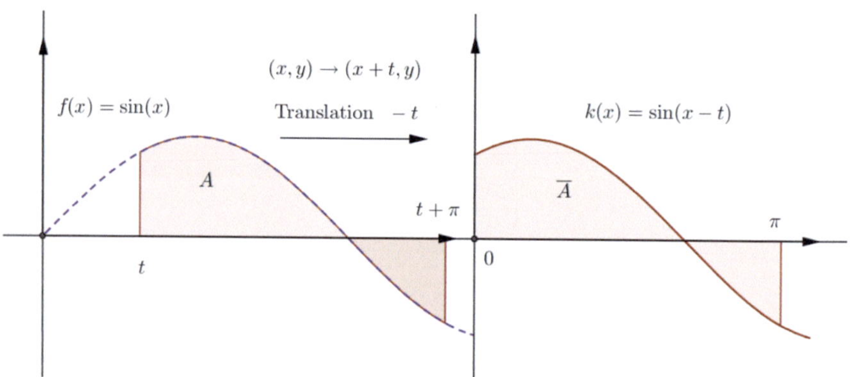

Fig. 3.28 Translation of the sine function

$$= \int_0^{\pi} (\sin(x)\cos(t) + \cos(x)\sin(t))dx$$

$$= \cos(t) \int_0^{\pi} \sin(x)dx + \sin(t) \int_0^{\pi} \cos(x)dx = C_S\cos(t),$$

because $\int_0^{\pi} \cos(x)dx = 0$ (see Fig. 3.27). Here again we put $\int_0^{\pi} \sin(x)dx = C_S$. From Fig. 3.26 we see that $\int_t^{t+\pi} \sin(x)dx = C_S - 2\int_0^t \sin(x)dx$ and thus $C_S \cos(t) = C_S - 2\int_0^t \sin(x)dx$ giving

$$\int_0^t \sin(x)dx = \frac{C_S(1 - \cos(t))}{2}.$$

We now want to determine the constant C_S. Again, we examine the area difference, which can be seen in Fig. 3.29 as a white triangle-like area, where we chose $t > 0$ but t close to 0.

We now want to estimate this area from above and below by triangle areas (see Fig. 3.30). The area under the sine function over the interval $[0, t]$=$[A, C]$ is on the one hand completely enclosed by the triangle *ACD*. On the other hand, it completely includes the triangle *ECB*, because the graph of the sine function lies between the line $y = x$ and the hypotenuse of the triangle *ECB*, where *EB* runs parallel to the line $y = x$.

This can be seen as follows: On the unit circle in Fig. 3.31, it is immediately apparent that for $0 \le x \le t \le \frac{\pi}{2}$ on the one hand $\sin(x) \le x$ and on the other hand $x - t + \sin(t) \le \sin(x)$, because.

$$t - x = \widehat{PQ} \ge \overline{PQ} \ge PR = \sin(t) - \sin(x).$$

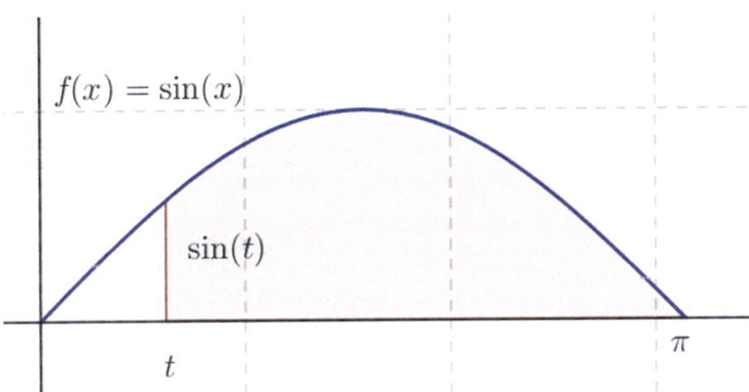

Fig. 3.29 Small triangle under the sine function

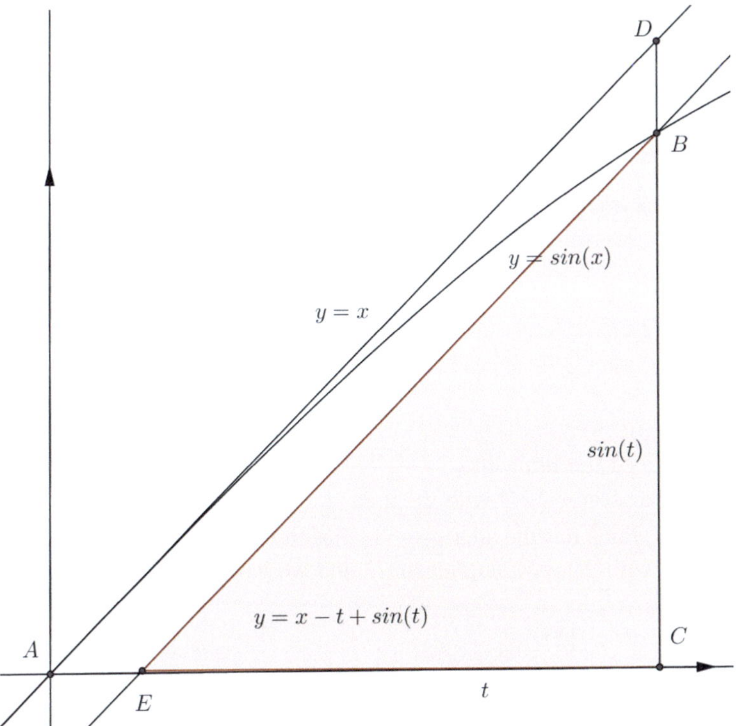

Fig. 3.30 Area under the sine function between two triangles

The difference angle $t - x$ as an arc is indeed longer than the segment PQ. This distance is the hypotenuse in the triangle PQR and thus longer than the cathetus $PR = \sin(t) - \sin(x)$, see Fig. 3.31.

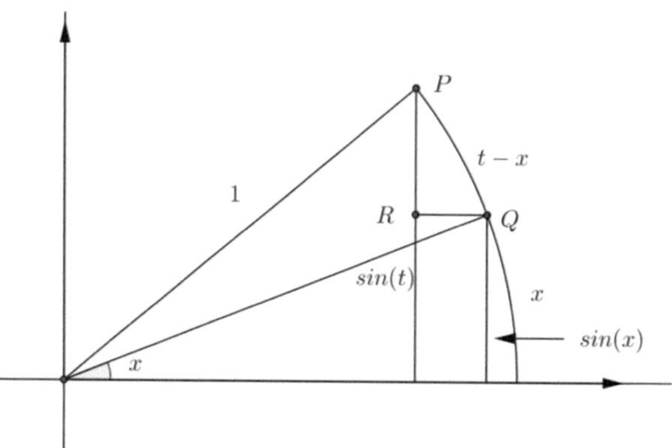

Fig. 3.31 The sine at the unit circle, part I

Now the triangle ACD has the area $t^2/2$, while the triangle ECB has the area $\frac{\sin^2 t}{2}$. Therefore, we have

$$\frac{\sin^2(t)}{2} \leq \frac{C_S(1-\cos(t))}{2} \leq \frac{t^2}{2} \text{ or } \sin^2(t) \leq C_S(1-\cos(t)) \leq t^2.$$

With this, we have

$$\frac{1-\cos^2(t)}{1-\cos(t)} = 1+\cos(t) \leq C_S \leq \frac{\sin^2(t)}{1-\cos(t)} \cdot \frac{t^2}{\sin^2(t)} = (1+\cos(t)) \cdot \frac{t^2}{\sin^2(t)}.$$

For this, we need the limit $\lim_{h\to 0} \frac{\sin(h)}{h}$. On the unit circle in Fig. 3.32, we see that the circular sector ABD with the area $\frac{t}{2\pi} \cdot \pi \cdot 1^2$ is entirely contained in the triangle ABF, which has the area $\frac{\sin t}{2\cos t} \cdot 1$. The circular sector, in turn, encloses the triangle AED with the area $\sin(t)\cos(t)/2$ and we have

$$\frac{\sin(t)\cos(t)}{2} \leq \frac{t}{2} \leq \frac{\sin(t)}{2\cos(t)} \text{ i.e., } \cos(t) \leq \frac{t}{\sin(t)} \leq \frac{1}{\cos(t)} \text{ and we have } \lim_{t\to 0} \frac{\sin(t)}{t} = 1.$$

The limit transition for $t \to 0$ then gives us $C_S = 2$. The final result is therefore

$$\int_0^t \sin(x)dx = \frac{C_S(1-\cos(t))}{2} = 1-\cos(t).$$

Of course, we can obtain the area under the cosine function in a similar way.

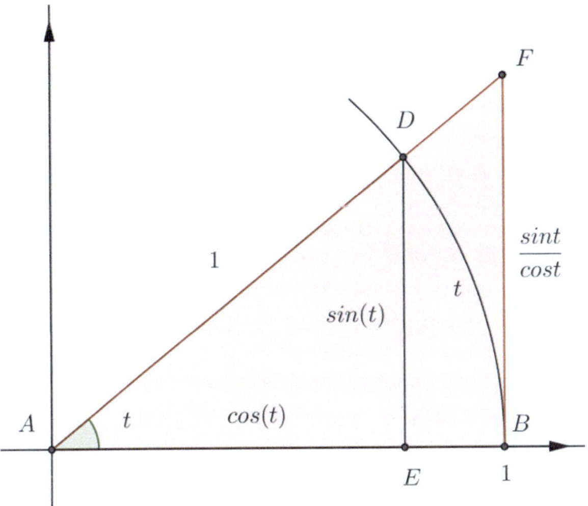

Fig. 3.32 The sine at the unit circle, part II

3.23 The Derivative of the Sine Function Through Calculus with Its Functional Property

Now we want to calculate the derivative of the sine function. At this point, the additive understanding of the infinitesimal increase fits best. For $f(x) = \sin(x)$ we have

$$\begin{aligned} f'(x) &= \lim_{h\to 0} \frac{f(x+h)-f(x)}{h} = \lim_{h\to 0} \frac{\sin(x+h)-\sin(x)}{h} \\ &= \lim_{h\to 0} \frac{\sin(x)\cos(h)+\cos(x)\sin(h)-\sin(x)}{h} \\ &= \sin(x)\cdot \lim_{h\to 0} \frac{\cos(h)-1}{h} + \cos(x)\cdot \lim_{h\to 0} \frac{\sin(h)-0}{h} \\ &= \sin(x)\cos'(0)+\cos(x)\sin'(0). \end{aligned}$$

So we are missing two derivative values at the point $x = 0$, namely $\sin'(0)$ and $\cos'(0)$. We now show that $\sin'(0) = 1$ and $\cos'(0) = 0$. Then the well-known formula

$$\sin'(x) = f'(x) = \sin'(0)\cos(x) + \sin(x)\cos'(0) = \cos(x)$$

follows. First, we show $\sin'(0) = 1$. In Sect. 3.22 above about the area under the sine curve, we had already determined the limit

$$\lim_{h\to 0} \frac{\sin(h)}{h} = 1$$

Upon closer inspection, we recognize that

$$1 = \lim_{h \to 0} \frac{\sin(h)}{h} = \lim_{h \to 0} \frac{\sin(h) - \sin(0)}{h} = \sin'(0).$$

Now we show $\cos'(0) = 0$. Indeed, we have

$$\begin{aligned}\cos'(0) &= \lim_{h \to 0} \frac{\cos(h) - 1}{h} = \lim_{h \to 0} \frac{\cos^2(h) - 1}{(\cos(h) + 1)h} \\ &= \lim_{h \to 0} \frac{\sin(h)}{\cos(h) + 1} \cdot \frac{\sin(h)}{h} = 0 \cdot 1 = 0\end{aligned}$$

So, the derivative of the cosine function at $x = 0$ is zero and we have determined the two missing derivative values.

3.24 The Derivative of the Sine Function with Translations and Scalings

Finally, we want to examine the slope of the tangent at the sine function $f(x) = \sin(x)$ in the second view with translations and scalings. At the point $(a, f(a))$ we consider the tangent with the slope $f'(a)$. This time we again perform a translation along the x-axis, where the shift vector again has the length t (Fig. 3.33).

In doing so, we obtain the function $k(x) = \sin(x - t)$. Because of the addition theorem, we know

$$k(x) = \sin(x - t) = \sin(x)\cos(t) - \sin(t)\cos(x).$$

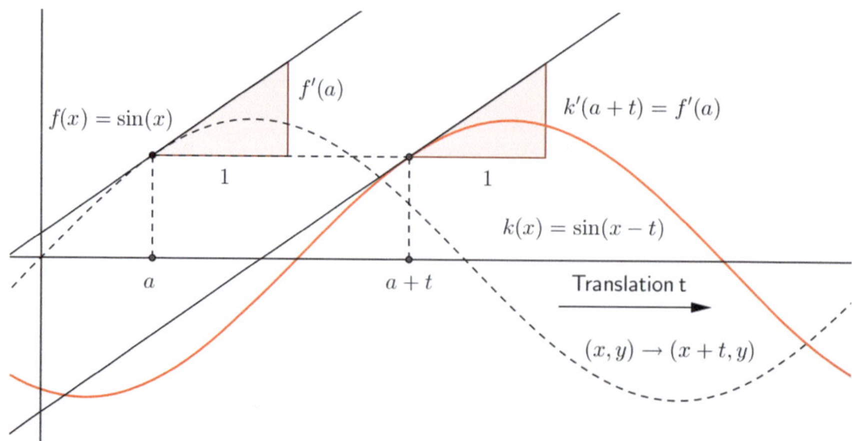

Fig. 3.33 Translation of the graph of the sine function

The translation of the function $f(x) = \sin(x)$ can be interpreted as a composition of a scaling with the factor $\cos(t)$ and the subtraction of another function, namely a scaled cosine function $\sin(t) \cdot \cos(x)$. For the derivative, we then have on one hand

$$f'(a) = \sin'(a) = k'(a+t),$$

because it is merely a translation, but on the other hand we have

$$k'(a+t) = \sin'(a+t)\cos(t) - \sin(t)\cos'(a+t),$$

because the translation can be interpreted as a scaling along with the subtraction of a scaled cosine function. Here we have used that the derivative of a sum of two functions is the sum of the corresponding derivatives. We set $t = -a$ and get

$$\sin'(a) = f'(a) = k'(a+t) = k'(0) = \sin'(0)\cos(t) - \sin(t)\cos'(0).$$

As above, we are still missing two derivative values $\sin'(0)$ and $\cos'(0)$, before we can state the final result. Above, we have calculated exactly these values, namely $\sin'(0) = 1$ and $\cos'(0) = 0$. Then the familiar formula follows

$$\sin'(x) = f'(x) = \sin'(0)\cos(x) + \sin(x)\cos'(0) = \cos(x).$$

The derivative of the cosine function can be treated similarly or transferred directly using a translation of the sine.

3.25 Bonus Example: The Logarithmic Spiral

It also turns out that the method presented can be applied to other functions and curves that are not covered in the school curriculum. If we consider the logarithmic spiral and replaces the transformations used above (translations and scalings) with a stretch $(x, y) \to (mx, my)$ and a rotation, then just like above, statements about areas and tangent slopes can be made. (A stretch can be seen as the combination of two scalings in x- and y-direction with the same scaling factor. The effect is an enlargement of all geometric elements.) It follows that we can also determine arc lengths and radii of curvature, which makes the logarithmic spiral special in this context (apart from the straight line). These observations go back to Jakob Bernoulli (Stillwell 1989, pp. 336–339), who worked extensively on the logarithmic spiral and called it *spira mirabilis*, or the wonderful spiral, due to its astonishing properties.

From now on, we use the definition of the logarithmic spiral in polar coordinates

$$r = f(\theta) = e^{k\theta},$$

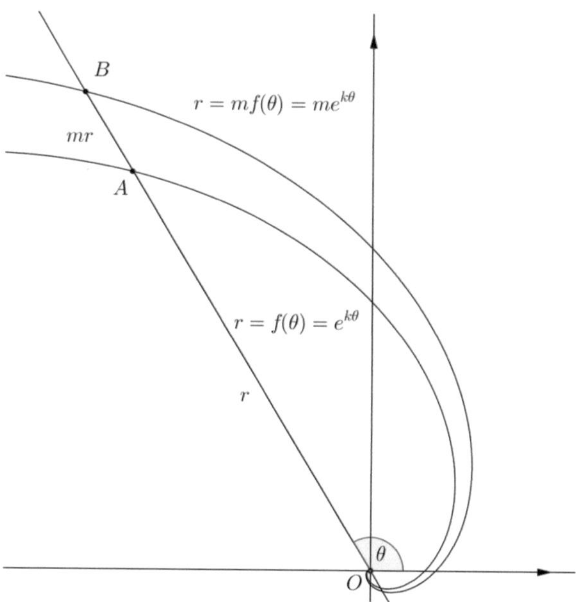

Fig. 3.34 The logarithmic spiral and a streched version of the logarithmic spiral

where k is a real constant. In the case of the stretch $(x, y) \to (mx, my)$ with stretch factor m, we then get

$$r = me^{k\theta} = e^{k\left(\theta + \frac{\ln(m)}{k}\right)} = e^{k(\theta+\alpha)}$$

and we see that we get the old spiral, where the angle θ is replaced by $\theta + \alpha$, which can be interpreted as the result of a rotation by the angle $\alpha = \ln(m)/k$ (see Fig. 3.34). We now want to use this "trick" to show some properties of the logarithmic spiral.

3.26 The Area Between a Radius Vector and the Logarithmic Spiral and the Calculation of the Arc Length

First, we consider the area between a radius vector and the logarithmic spiral. In the case of the stretch with stretch factor m, this area increases to the m^2-fold of the original area. The subsequent rotation does not change the area, of course. Therefore, we get a quadratic formula

$$P = Qr^2$$

for the area P, where the constant Q indicates the area for $r = 1$ (see Fig. 3.35).

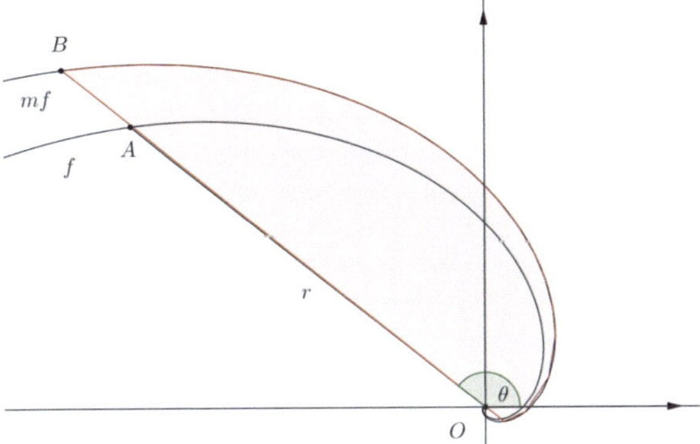

Fig. 3.35 The area between the logarithmic spiral and a radius vector

Surprisingly, the same argument can also be applied to the arc length of the logarithmic spiral. In the case of the stretch, the arc length increases to the m-fold of the original arc length, and the subsequent rotation does not change the length (see Fig. 3.35). Overall, we thus get a linear relationship between the length of the radius vector r and the arc length B of the spiral,

$$B = Gr,$$

where G is the length of the arc for $r = 1$. For the determination of the various constants, see Stillwell (1989, pp. 336–339).

3.27 The Tangent Slope at the Logarithmic Spiral

Now we consider a tangent to the original spiral at the point B. In the case of the stretch, angles are preserved, so the angle between the radius vector and the tangent at the new and the old curve are the same. The subsequent rotation does not change the angle between the radius vector and the tangent, so we can immediately conclude that the angle between the radius vector and the tangent along the entire spiral is constant (see Fig. 3.36). This then solves the problem of the tangent slope, i.e., the derivative.

The special properties of the functions or curves considered here, namely, that the combination of two transformations leads us back to the original curve, makes it possible to calculate tangent slopes and integrals (and in the case of the logarithmic spiral also arc lengths) relatively easily.

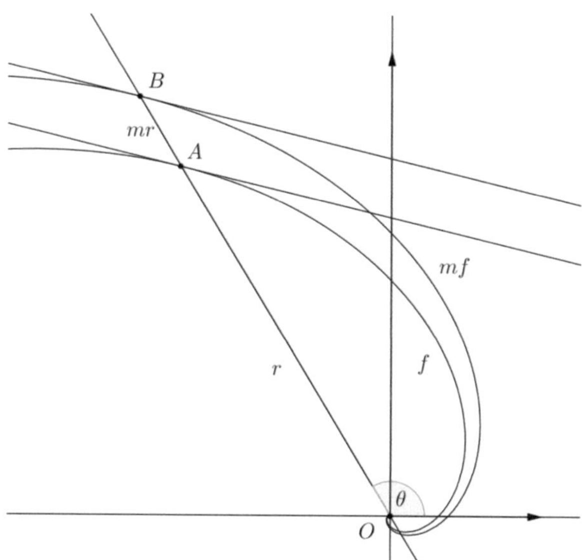

Fig. 3.36 The tangent at the logarithmic spiral

3.28 Summary

With the use of word processing or drawing programs and the viewing, sharing, and editing of images on digital tools, enlargements, scalings and translations have found a new role in the everyday life of pupils, students and scholars (see Fig. 3.37). These new everyday experiences may possibly make it easier for students and pupils to access the infinitesimal calculus presented here.

With the "Method of Gregorius" supported by scalings and translations, we can geometrically integrate not every continuous function per se, but all elementary functions that occur in school mathematics.

The use of the method presented here requires only little prior knowledge of the students.

- In the case of the power function, they must know the relationship $(tx)^k = t^k \cdot x^k$.
- In the case of the logarithm, they must know the relationship $\ln(tx) = \ln(x) + \ln(t)$.
- In the case of the exponential function, they must know the relationship $g^{x+t} = g^x \cdot g^t$.
- In the case of the sine function, they must know the addition theorem $\sin(x+t) = \sin(x)\cos(t) + \cos(x)\sin(t)$.

These are the fundamental properties of the mentioned functions and are often known to students before they tackle integral calculus. In school, the examination of these properties of specific functions plays a central role.

Fig. 3.37 Gregorius of St. Vincent scaled and unscaled

Differentiation can also be understood through the special functional properties of functions. We have demonstrated this for each type of function.

With this approach, the concept of derivative could be introduced after the integration and initially without the concept of tangent. If we want to differentiate a function of the form $F(x) = bx^{n+1}$, we first recognize it as the area function under the graph of $f(x) = (n+1)bx^n$, where f is obtained by "anti-integration" of F. Then $F(x+h) - F(x)$ is the area of a strip of width h, whose height is between $f(x)$ and $f(x+h)$. For monotonic functions this means

$$f(x) \leq \frac{F(x+h) - F(x)}{h} \leq f(x+h) \text{ or } F' = f.$$

In short, this sheds new light on the main theorem of differential and integral calculus.

Remark 3.3
In previous sections, we have seen that we can determine the areas under, for example, the power functions if we know them over a certain interval [0,1]. This also applied to other types of functions such as logarithm, exponential function, and trigonometric functions. We were also able to determine the derivatives of these functions as soon as we could specify the tangent slopes at a single point of the function. In order to calculate these individual area values or tangent slopes, we had to perform specific limit considerations at individual points. At this point, we would like to once again precisely identify where this was the case, to show how few real limit calculations are necessary to operate the analysis of all so-called school functions.

- *For the logarithm and exponential function, we naturally needed the definition of the Euler constant* $e = \lim_{n\to\infty}\left(1 + \frac{1}{n}\right)^n$

- *We also needed the derivative of the logarithm function at* $x = 1$, *i.e.,* $\lim_{h\to 0} \frac{L(1+h)-L(1)}{h} = 1$.
- *For the power functions, we needed* $\lim_{r\to 1} \frac{r^n-1}{r-1} = n$.
- *For the exponential function, we encountered* $\lim_{h\to 0} \frac{e^h-1}{h} = 1$.
- *For the trigonometric functions, we needed* $\sin'(0) = \lim_{h\to 0} \frac{\sin(h)}{h} = 1$ *and* $\cos'(0) = \lim_{h\to 0} \frac{\cos(h)-1}{h} = 0$.

These are the only six limits that we actually needed at specific points (zero or one) to cover the entirety of all school functions.

These considerations connect a range of different subject areas in the broader field of geometry, algebra, and analysis. But what is the purpose of these considerations? This contribution is intended for use in high school education. The extent to which a consistent conceptual development of differential and integral calculus based on transformation geometry can be implemented is open for discussion and is also a matter of taste.

This approach can exemplify that an alternative to the traditional way of computing the limit of the difference quotient is possible. The approach proposed here is based on currently intensively taught subjects such as function graphs or functional properties and also includes geometric aspects such as transformations of the plane and the coordinate system in the calculations of the difference quotients. The functional properties become the central tool in this presentation to solve concrete problems, such as calculating the derivative or determining areas.

But in addition to high school students who are further interested in analysis, we primarily think of possible uses in mathematics teacher education. In our experience, many prospective teachers only move on one well-trodden path into the garden of analysis and can only further develop the contributions of students in analysis lessons if they are precisely classified on this path. We believe it is important to provide prospective teachers with a variety of approaches and perspectives in analysis, to which we want to contribute here. Our considerations above are by no means intended to replace the classic approaches to analysis, but to supplement them.

Also, as in Kaenders (2014), we use the base functions in a way that highlights their character as individuals among the functions. Historically, there were initially such individual functions before a theory of all continuously differentiable functions emerged. We believe that in teaching, a deeper examination of these important basic functions should precede a general treatment of the entire class of continuously differentiable functions. Therefore, we always refer back to the functional properties of these base functions as far as possible.

In this chapter, we have shown that the presented method for area calculation and integration through simple geometric transformations can be seamlessly transferred to differentiation and the determination of tangent slopes. The method thus has a greater range than we might initially suspect. However, it must also be noted that a transfer to other phenomena of analysis, such as curve lengths or radii of curvature, is not easily achievable because these quantities are not always compatible with scalings.

The approach pursued here leads to the same insights as the classic considerations. However, this approach allows for a different, more geometric understanding of infinitesimal calculus in teacher education and school. Our considerations not only lead to the known formulas for derivatives and integrals but also expand the variety of methods and perspectives. They are certainly not intended to replace the classic approaches. With this approach, the scope of methodological ideas, such as working with transformation geometry in determining the derivative and integral of functions in the school context, becomes visible.

In calculating the constants in the case of area calculation, we considered small area differences and then the associated limits. This technique already points to the main theorem of differential and integral calculus and can later be picked up and generalized into a full proof of this theorem.

We did not further address the existence of the tangent or the area under the function, i.e., the question of differentiability or integrability, in this chapter. We assume these and then show how knowledge of a tangent slope at one point of a function can gain knowledge about the tangent slope at all other points of the same function, or how knowledge of one area can gain knowledge about all such areas. This can be interpreted as a weakness of our presentation because we might want to address the difficult question of general differentiability or integrability in class. On the other hand, the presentation becomes geometrically more vivid and builds on already familiar ideas—if at least a basic geometric education is present. Thus, limit considerations—at least in the purely geometric part of the presentation—play a subordinate role. The method shown here is therefore probably particularly suitable within the framework of an introduction to analysis.

Some of our ideas appear in the literature by authors who approach derivatives and areas under functions in a similar way, but do not cover all school functions as comprehensively. Freund (1960), Kirsch (1960), Püschel (1960), and Freudenthal (1973) have also highlighted and used comparable transformation properties of individual functions. Freund (1960) and Freudenthal (1973) use these properties to calculate areas under power functions with natural exponents, which they can initially only determine up to a multiplicative constant. Freund (1960) then conducts a complex proof to determine this constant using the Bernoulli inequality. Püschel (1960) only calculates the areas under functions $y = 1/x^m$ for $m \geq 2$, by reducing them to improper integrals.

Bibliography

Bopp, K. (1907): Die Kegelschnitte des Gregorius a St. Vincentino in vergleichender Bearbeitung, in: Abhandlung zur Geschichte der mathematischen Wissenschaften mit Einschluss ihrer Anwendungen, Vol. 20, 2.Part.

Burn, R. P. (1999): *Integration, a genetic introduction,* Nordisk matematikkdidaktikk, Nr. (7)1.

Burn, R. P. (2001) *Alphonse Antonio de Sarasa and logarithms*, Historia mathematica, Vol. 28 (1), pp. 1–17.

Edwards, C. H. (1979): *The Historical Development of the Calculus.* Springer Heidelberg.

Freund, H. (1960): Die Gewinnung von Steigungswerten durch analytisch-geometrische Betrachtungen. In: Freund, H. (Hrsg.) Geometrische Hilfsmittel in der Analysis. Der Mathematikunterricht, 6(2), Friedrich-Verlag, Seelze, pp. 22–51.

Freudenthal, H. (1973): Mathematik als pädagogische Aufgabe, Vol. 1., Klett Stuttgart.

Gregorius a San Vincentio (1647). Problema austriacum plus ultra quadratura circuli. Antwerpen, Mersius. (Opus geometricum quadratura circuli et sectionum coni).

Kaenders, R. H., (2014): *Von einem kognitiven Konflikt zur Quadratur der Parabel.* Beiträge zum Mathematikunterricht, GDM-Tagungsbericht 2014, pp. 583–586.

Kaenders, R. H., (2015) *Flächenbestimmung mit Ähnlichkeit als Alternative zur sogenannen "h-Methode"*, Beiträge zum Mathematikunterricht, GDM-Tagungsbericht 2015, pp. 440–443.

Kaenders, R. & Kirfel, C. (2016): Weiterentwicklung historischer Zugänge zur Infinitesimalrechnung über Elementargeometrie, Beiträge zum Mathematikunterricht 2016, hrsg. v. Institut für Mathematik und Informatik der Pädagogischen Hochschule Heidelberg. Münster WTM-Verlag, pp. 1235–1238.

Kaenders, R. & Kirfel, C. (2017): Flächenbestimmung bei Basisfunktionen der Schule mit Elementargeometrie. Math. Semesterberichte (2017). https://doi.org/10.1007/s00591-017-0191-6

Kaenders, R. & Kirfel, C. (2020a): Ableitung und Integral bei Basisfunktionen der Schule mit Elementargeometrie (Teil I), MNU, Jahrgang 73, 2/2020, pp. 156–162.

Kaenders, R. & Kirfel, C. (2020b): Ableitung und Integral bei Basisfunktionen der Schule mit Elementargeometrie (Teil II), MNU, Jahrgang 73, 4/2020, pp. 328–333.

Kirfel, C. (2014). *Integration by geometrical means—a unified approach.* Mathematics Teaching 239. pp. 23–25.

Kirfel, C. (2018): Die logarithmische Spirale – ein dankbares Studienobjekt, Beiträge zum Mathematikunterricht 2018, Gemeinsame Jahrestagung der GDM und der DMV Paderborn, März 2018, Münster WTM-Verlag, pp. 955–958.

Kirsch, A. (1960): Ein geometrischer Zugang zu den Grundbegriffen der Differentialrechnung. In: Freund, H. (Hrsg.) Geometrische Hilfsmittel in der Analysis. Der Mathematikunterricht, 6(2), Friedrich-Verlag, Seelze, pp. 5–21. (1960).

Püschel, W. (1960): Direkte Integrationsmethoden von einigen Klassen von Potenzfunktionen mit rationaler Hochzahl. In Freund, H. (Hrsg.) Geometrische Hilfsmittel in der Analysis. Der Mathematikunterricht, 6(2), Friedrich-Verlag, Seelze, pp. 52–65.

Stillwell, J. (1989): *Mathematics and Its History,* Springer-Verlag.

Toeplitz, O. (1949): Die Entwicklung der Infinitesimalrechnung – eine Einführung in die Infinitesimalrechnung nach der genetischen Methode, Springer-Verlag.

Van den Broek, L. (1994): De afgeleide van meetkundig afgeleid, Euclides, 70(1), pp. 7–10.

Volkert, K. (1996): Die Quadratur der Hyperbel des Gregorius a San Vincentio. Herrn Professor Dr. U. Loettgen (Koeln) zum 65. Geburtstag gewidmet. Journal für Mathematik-Didaktik 17 (1), pp. 3–20.

Two Methods of Integration by Fermat

4

4.1 Introduction

When integrals are introduced in modern mathematics education, it often happens in various ways. Processes of change and accumulation are widespread in many representations of the topic of integration in school, see Roth and Siller (2016). However, the traditional calculation of areas under curves is still an approach to the integral that has been preserved in schools and universities. This often relies on calculations using calculators and computer programs. In some cases (e.g., Brand et al. 2011 and Körner et al. 2011), formulas for the integral of simple power functions are still traditionally calculated using infinite sums. Here, the first Fermat method with equidistant intervals is used. However, this often only happens in exercises in textbooks and is no longer core material. In many of the modern representations, integration is then only "antidifferentiation", i.e., the reverse of differentiation. In this case, the initial geometric task of finding an area is reinterpreted as an algebraic algorithmic task. With such an algorithmic perspective, the illustrative, geometrically intuitive understanding is lost.

In this chapter, we would like to present and analyze the integration methods that Fermat (1601–1665) developed for power functions. It turns out that the second Fermat method, in particular, has great potential, which can also be used to tackle other functions that Fermat himself was not able to tackle. The first method can occasionally still be found in textbooks in school or university. In contrast, the second method has been forgotten and is mainly found in historical texts.

The main source for this chapter is Fermat's treatise with the detailed title.

> "Sur la transformation et la simplification des équations de lieux, pour la comparaison sous toutes les formes des aires curvilignes, soit entres elles, soit avec des rectilignes, et en même temps sur l'emploi de la progression géométrique pour la quadrature des paraboles et hyperboles à l'infini." (Fermat 1896).

© The Author(s), under exclusive license to Springer-Verlag GmbH, DE, part of Springer Nature 2026

C. Kirfel, *Side Paths in the History of Mathematics*, Mathematics Study Resources 21,
https://doi.org/10.1007/978-3-662-72918-2_4

Part of the material can also be found in "Fermat's Methoden zur Integration" (Kirfel 2017A) and in Robert Burns' article, "Integration a genetic Introduction" (Burn 1999). Burn goes further in the development of the concept of the integral and reaches up to the Lesbeque integral. However, Burn does not include the explanations for the integral of the logarithm. Our concern is to compare the two Fermat methods and uncover the still unrealized potential of these methods.

4.2 Fermat's First Method

In the first Fermat integration method, equidistant rectangles are used, i.e., rectangles that are based on an interval division with equal-length subintervals. All rectangles have the same "width". This method works as follows. We want to calculate the area under the graph of a power function $f(x) = x^k$ for example over an interval between $x = 0$ and $x = a$. Fermat divided the interval $[0, a]$ into n equal-length subintervals of the same length $\frac{a}{n}$.

Over each of these intervals, he then draws two rectangles, one that just reaches the function graph and one that reaches up to the highest point of the graph in this interval. Now the areas of the rectangles under the function graph are summed up and we obtain a lower bound for the area under the function graph. Correspondingly, we also obtain an upper bound for this area if we sum up the areas of the rectangles that extend beyond the function graph. The sum of the areas of the rectangles above the function graph gives the upper sum, those below the function graph the lower sum, as indicated in Fig. 4.1. In the case of $y = x^2$ this is calculated as follows:

Upper sum:

$$\begin{aligned} O &= \frac{a}{n} \cdot \left(\frac{a}{n}\right)^2 + \frac{a}{n} \cdot \left(\frac{2a}{n}\right)^2 + \frac{a}{n} \cdot \left(\frac{3a}{n}\right)^2 \cdots + \frac{a}{n} \cdot \left(\frac{na}{n}\right)^2 \\ &= \left(\frac{a}{n}\right)^3 \cdot \left(1^2 + 2^2 + 3^3 + \cdots + n^2\right) \end{aligned}$$

Lower sum:

$$\begin{aligned} U &= \frac{a}{n} \cdot 0 + \frac{a}{n} \cdot \left(\frac{a}{n}\right)^2 + \frac{a}{n} \cdot \left(\frac{2a}{n}\right)^2 + \cdots + \frac{a}{n} \cdot \left(\frac{(n-1)a}{n}\right)^2 \\ &= \left(\frac{a}{n}\right)^3 \cdot \left(1^2 + 2^2 + 3^3 + \cdots + (n-1)^2\right) \end{aligned}$$

We now use the formula $1^2 + 2^2 + 3^3 + \ldots + n^2 = \frac{2n^3+3n^2+n}{6}$, which we will derive later. After some transformations, the following inequality for the area A can be worked out

$$\frac{a^3}{3} \cdot \frac{(2n^3 - 3n^2 + n)}{2 \cdot n^3} < A < \frac{a^3}{3} \cdot \frac{(2n^3 + 3n^2 + n)}{2 \cdot n^3}.$$

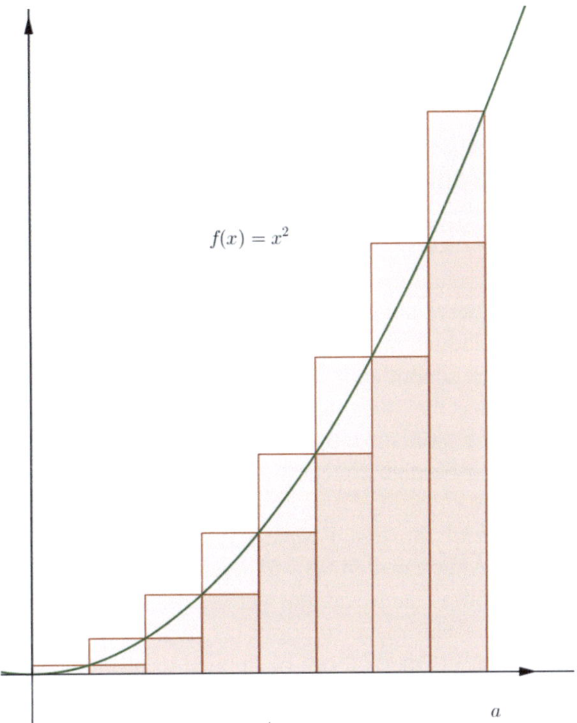

Fig. 4.1 Upper and lower sum

Now these inequalities should be valid for all values of n and thus only one value for A is possible, namely

$$A = \frac{a^3}{3}.$$

Here we have applied the formula $1^2 + 2^2 + 3^2 + \cdots + n^2 = \frac{n(n+1)(2n+1)}{6}$, without specifying where it comes from or how it can be proven. A natural method of proof here is induction. The formula is valid for $n = 1$, because due to $1^2 = 1$ and $\frac{1 \cdot 2 \cdot (2+1)}{6} = 1$, so both sides of the equation match. Now, let's assume that the formula is correct for a value k, so

$1^2 + 2^2 + 3^2 + \cdots + k^2 = \frac{k(k+1)(2k+1)}{6}$,

then

$$\begin{aligned} 1^2 + 2^2 + 3^2 + \cdots + k^2 + (k+1)^2 &= \frac{k(k+1)(2k+1)}{6} + (k+1)^2 \\ &= \frac{(k+1)(k(2k+1) + 6(k+1))}{6} \\ &= \frac{(k+1)(k+2)(2k+3)}{6}. \end{aligned}$$

Thus, the formula also holds for $k+1$ and by the principle of induction, for all natural numbers n.

Let's examine the function $f(x) = x^3$ in the same way and use the formula $1^3 + 2^3 + 3^3 + \cdots + n^3 = \frac{n^2(n+1)^2}{4}$, it follows that the area under the graph of a third-degree power function over the interval $[0, a]$ must be equal to $A = a^4/4$.

With appropriate means, the method can be extended to cover all power functions $y = x^k$ where the exponent is a natural number, and we obtain $\int\limits_0^a x^k dx = \frac{a^{k+1}}{k+1}$, provided we know a formula for $1^k + 2^k + 3^k + \ldots + n^k$. For the sum of the first n square numbers, we were able to derive the corresponding summation formula above with the help of induction. In the general case, such summation formulas are not easy to derive in the school context. The sum formulas generally contain very complicated expressions, the so-called Bernoulli numbers. In addition, the given method only works for exponents k that are natural numbers. These are shortcomings inherent to the first Fermat's integration method, and we would like to avoid them.

Another application of the first Fermat's integration method is found in the exponential function. Although Fermat was familiar with the logarithm function, the exponential function as such was not yet known during Fermat's lifetime. It may therefore sound anachronistic to apply his method to this function. On the other hand, the fact that his method can be applied to phenomena unknown at his time, demonstrates how powerful the method is.

To calculate the integral under the exponential function over the interval $[0, a]$, we divide the interval again into n equal parts of length $h = a/n$, so $a = n \cdot h$. Thus, we chose an equidistant interval division. Then we can write down the lower sum of the Riemann integral as follows (see Fig. 4.2):

$$U = \sum_{k=0}^{n-1} h \cdot f(k \cdot h) = h \sum_{k=0}^{n-1} e^{kh} = h \sum_{k=0}^{n-1} \left(e^h\right)^k = h \frac{e^{nh} - 1}{e^h - 1} = \left(e^a - 1\right) \frac{h}{e^h - 1}.$$

Here we have used the formula for finite geometric series. For the integral I, or the limit, we have

$$I = \left(e^a - 1\right) \lim_{h \to 0} \frac{h}{e^h - 1} = \frac{e^a - 1}{G},$$

where we have set

$$G = \lim_{h \to 0} \frac{e^h - 1}{h}.$$

Here we only used the property $e^{s+t} = e^s \cdot e^t$. Upon closer inspection, we notice that G is just the derivative of the exponential function for $x = 0$. We have already determined this derivative in Chap. 3 on the method of Gregorius. There we found $G = 1$ and thus $I = e^a - 1$. The integral of the exponential function is therefore

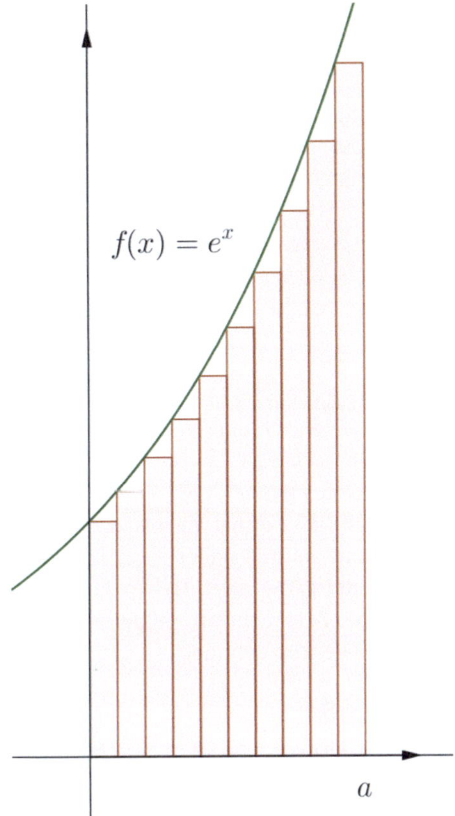

Fig. 4.2 Lower sum for the exponential function

again an exponential function. The reduction -1 represents the area under the curve from $-\infty$ to zero.

Interestingly, the same limit G also appears in the calculation of the derivative of the exponential function at any point x, not just at $x = 0$. We have

$$f'(x) = \lim_{h \to 0} \frac{f(x+h) - f(x)}{h} = \lim_{h \to 0} \frac{e^{x+h} - e^x}{h} = e^x \lim_{h \to 0} \frac{e^h - 1}{h} = e^x \cdot G.$$

In this way, the connection between derivative and area calculation for the exponential function becomes particularly clear.

Riemann et al. (1876) emphasizes in his definition of the so-called Riemann integral that the limits of the upper and lower sums must be equal, for all conceivable refinements of the interval division, where the maximum subinterval width approaches zero. Fermat only considers equidistant interval divisions in the first method. Why this is sufficient here is often not a central topic in school.

4.3 Fermat's Second Method

In his treatise "Sur la transformation" (Fermat 1896), Fermat proposes a second method that does not have the disadvantages mentioned above (see also Burn 1999), namely the problem of finding a corresponding summation formula for the associated powers for each new power function. The main difference is that Fermat does not choose an equidistant division of the interval $[0, a]$, but allows the subintervals, or the abscissa values of the endpoints of the interval division, to form a geometric sequence. He borrowed this idea from Archimedes, as he himself writes (Fermat 1896, p.1, author's translation).

> "Whatever the case may be, I value this sequence and have found that it is very fruitful for quadratures, and I am happy to share my discovery with any modern geometer. This discovery allows us to square both parabolas and hyperbolas using exactly the same method."

By "squaring," Fermat means determining the area. Originally, the problem of determining an area under a curve was described as the task of finding a square equal to the given area. By "parabolas," he means power functions $y = x^k$ for natural numbers k and by hyperbolas, he means functions $y = 1/x^k$, again with natural k. Fermat continues:

> "This method is derived from a single well-known property of the geometric series, namely the following theorem: Given a geometric series whose terms continuously decrease. The difference between two consecutive terms of this series is to the smallest of the two as the largest of all terms to the sum of all the other terms up to infinity." (ibid., p. 1).

With modern notation, Fermat thus claims the following for a geometric series with the first term a and factor q:

$$\frac{a - aq}{aq} = \frac{aq - aq^2}{aq^2} = \cdots = \frac{a}{aq + aq^2 + aq^3 + \cdots}, \tag{4.1}$$

thus

$$a + aq + aq^2 + \cdots = a + \frac{a \cdot aq}{a - aq} = \frac{a}{1 - q},$$

as we know it today (see Fig. 4.3). Fermat further writes (ibid. p.2):

> "I now claim that all these infinitely many hyperbolas, with the exception of that of Apolonius, or the primary hyperbola $\left(y = \frac{1}{x}\right)$ can be squared using the geometric series with the same method."

Using the example of the hyperbola $y = \frac{1}{x^2}$, he shows how his method works. The distances $AG, AH, AO, \ldots$ form a geometric sequence. In modern notation,

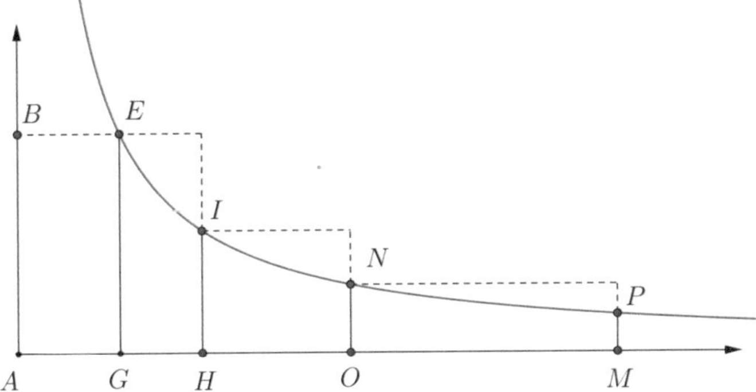

Fig. 4.3 Fermat's illustration

this could be written as $AH = AG/q$, $AO = AG/q^2$ etc., where the common factor must be $q < 1$.

Fermat now shows that the areas of the associated rectangles (only the diagonals are given) $EH, IO, NM, \ldots$ also form a geometric sequence, the sum of which he then calculates later. We first see

$$\frac{\text{Area}(IO)}{\text{Area}(EH)} = \frac{HO \cdot HI}{GH \cdot GE} = \frac{HO \cdot \left(\frac{1}{AH}\right)^2}{GH \cdot \left(\frac{1}{AG}\right)^2} = \frac{(AO - AH)AG^2}{(AH - AG)AH^2} = \frac{\left(\frac{AG}{q^2} - \frac{AG}{q}\right)AG^2}{\left(\frac{AG}{q} - AG\right)AH^2}$$

$$= \frac{AG/q\left(\frac{1}{q} - 1\right)AG^2}{AG\left(\frac{1}{q} - 1\right)\left(\frac{AG}{q}\right)^2} = q.$$

We get the same ratio for the subsequent rectangles. Using the above formula (4.1), Fermat can now give the result of the total rectangle sum S (to the right of GE). We use the letter R for the total sum excluding the first term of S. The ratio of the difference of two neighboring terms to the smaller of the two is equal to the ratio of the largest of all terms to the sum of the remaining terms, thus R.

$$\frac{GH \cdot GE - HO \cdot HI}{HO \cdot HI} = \frac{GH \cdot GE}{R}$$

or

$$S = R + GH \cdot GE = GH \cdot GE + \frac{HG \cdot GE \cdot HO \cdot HI}{HG \cdot GE - HO \cdot HI}.$$

After some further transformations, Fermat obtains

$$S = GH \cdot GE + \frac{AG}{AG^2} = \text{Area}(BH).$$

Fermat then makes some remarks about the infinitesimal transition. Afterwards, he obtains as a result that the area under the curve to the right of GE is equal to the rectangle BH, which in the limit case is equal to BG (see Fig. 4.3). To name the rectangles, Fermat again only gives the diagonals. He then shows how the method can be used with minor changes for higher hyperbolas, i.e., power functions $f(x) = \frac{1}{x^k}$ with integer exponents $k \geq 2$.

We would like to present the method here using modern terminology and calculate the area over the interval $[a, \infty]$. Fermat uses the interval subdivision: $a, ar, ar^2, ar^3, \ldots, ar^n, \ldots$ for $r > 1$, so that the points form an infinitely increasing geometric sequence (see Fig. 4.4). We assume here $a > 0$.

Initially, we do not know whether the area is finite at all. However, we will find an upper bound for the area, which means the area must be finite.

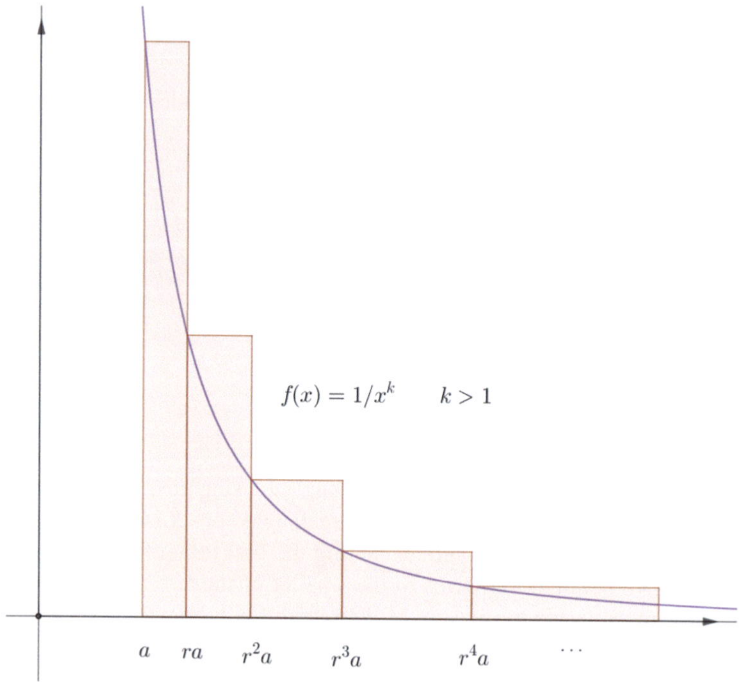

Fig. 4.4 Division of the interval for the higher hyperbola

The upper sum O is here

$$\begin{aligned}O &= a(r-1)\cdot\frac{1}{a^k} + a\left(r^2-r\right)\cdot\left(\frac{1}{ra}\right)^k + a\left(r^3-r^2\right)\cdot\left(\frac{1}{r^2a}\right)^k + \cdots\\ &\quad + a\left(r^{n+1}-r^n\right)\cdot\left(\frac{1}{r^na}\right)^k + \cdots\\ &= \frac{r-1}{a^{k-1}}\left(1+\frac{1}{r^{k-1}}+\frac{1}{r^{2(k-1)}}+\frac{1}{r^{3(k-1)}}+\cdots+\frac{1}{r^{n(k-1)}}+\cdots\right)\\ &= \frac{r-1}{a^{k-1}}\cdot\frac{1}{1-\frac{1}{r^{k-1}}} = \frac{r-1}{a^{k-1}}\cdot\frac{r^{k-1}}{r^{k-1}-1}.\end{aligned}$$

Here we were able to use the formula for infinite geometric series because of $0 < \frac{1}{r^{k-1}} < 1$.

For the lower sum, we get a corresponding result.

$$\begin{aligned}U &= a(r-1)\cdot\left(\frac{1}{ra}\right)^k + a\left(r^2-r\right)\cdot\left(\frac{1}{r^2a}\right)^k + \cdots\\ &\quad + a\left(r^n-r^{n-1}\right)\cdot\left(\frac{1}{r^na}\right)^k + \cdots\\ &= \frac{r-1}{r^ka^{k-1}}\left(1+\frac{1}{r^{k-1}}+\frac{1}{r^{2(k-1)}}+\frac{1}{r^{3(k-1)}}+\cdots+\frac{1}{r^{n(k-1)}}+\cdots\right)\\ &= \frac{r-1}{r^ka^{k-1}}\cdot\frac{1}{1-\frac{1}{r^{k-1}}} = \frac{r-1}{ra^{k-1}}\cdot\frac{1}{r^{k-1}-1}.\end{aligned}$$

The exponent $k-1$ is a natural number. Therefore, we can use the formula for finite geometric series and get for the upper and lower sum:

$$\begin{aligned}O &= \frac{r^{k-1}}{a^{k-1}}\cdot\frac{r-1}{r^{k-1}-1} = \frac{r^{k-1}}{a^{k-1}}\cdot\frac{1}{1+r+r^2+\cdots+r^{k-2}}\\ U &= \frac{1}{ra^{k-1}}\cdot\frac{r-1}{r^{k-1}-1} = \frac{1}{ra^{k-1}}\cdot\frac{1}{1+r+r^2+\cdots+r^{k-2}}\end{aligned}$$

The limit for $r\to 1$ is now easy to calculate, because $\lim_{r\to 1} r^{k-1} = \lim_{r\to 1} r = 1$ and $\lim_{r\to 1}\left(1+r+r^2+\cdots+r^{k-2}\right) = k-1$. In total, we get for the area

$$\begin{aligned}U &= \frac{1}{ra^{k-1}}\cdot\frac{1}{1+r+r^2+\cdots+r^{k-2}} < A < \frac{r^{k-1}}{a^{k-1}}\cdot\frac{1}{1+r+r^2+\cdots+r^{k-2}}\\ &= O\end{aligned}$$

thus

$$A = \frac{1}{a^{k-1}(k-1)}.$$

This describes the area from a to infinity.

If the exponent k is not a natural number, we can no longer use the formula

$$r^{k-1} - 1 = (r-1)\left(1 + r + r^2 + \ldots + r^{k-2}\right)$$

This deficiency can, however, be remedied, because

$$\lim_{r \to 1} \frac{r-1}{r^{k-1}-1} = \lim_{r \to 1} \frac{1}{(k-1)r^{k-2}} = \frac{1}{k-1},$$

using L'Hospital's rule, which applies to all values of k, including fractions and irrational numbers, if the corresponding derivative formulas have already been shown.

4.4 Higher Parabolas

Fermat refers to the power functions $f(x) = x^k$ with $k \geq 2$ as higher parabolas, as mentioned above. We now examine the area from 0 to a instead of from a to infinity.

Fermat's method is now presented using the example function $f(x) = x^2$: Fermat chooses a constant $0 < r < 1$ and starts at the upper end a of the interval and lets the division points then form a geometric sequence with factor $r < 1$. So, he chooses the sequence:$\ldots, ar^n, ar^{n-1}, \ldots ar^3, ar^2, ar, a$ for the division of the interval (see Fig. 4.5).

This subdivision contains infinitely many subintervals. Again, he erects rectangles over these intervals and gets in a very similar way as above an upper and a lower sum, which give him an upper and a lower bound for the area under the function graph. Unlike the first Fermat method, where Fermat initially considered finite sums of rectangles, this sum is already an infinite sum of rectangles, which then represents an upper sum, because each rectangle encompasses the associated curve area.

For the upper sum, the following applies:

$$\begin{aligned} O &= a(1-r) \cdot a^2 + a\left(r - r^2\right) \cdot (ra)^2 + a\left(r^2 - r^3\right) \cdot \left(r^2 a\right)^2 + \cdots \\ &\quad + a\left(r^n - r^{n+1}\right) \cdot \left(r^n a\right)^2 + \cdots = a^3(1-r) \\ &\quad \cdot \left(1 + r^3 + r^6 + r^9 + \cdots + r^{3n} + \cdots\right) = \frac{a^3(1-r)}{1-r^3}. \end{aligned}$$

Here we have used the formula for the infinite geometric series with $0 < r < 1$. But now $1-r^3 = (1-r)\left(1 + r + r^2\right)$, which can be easily be shown by expanding,

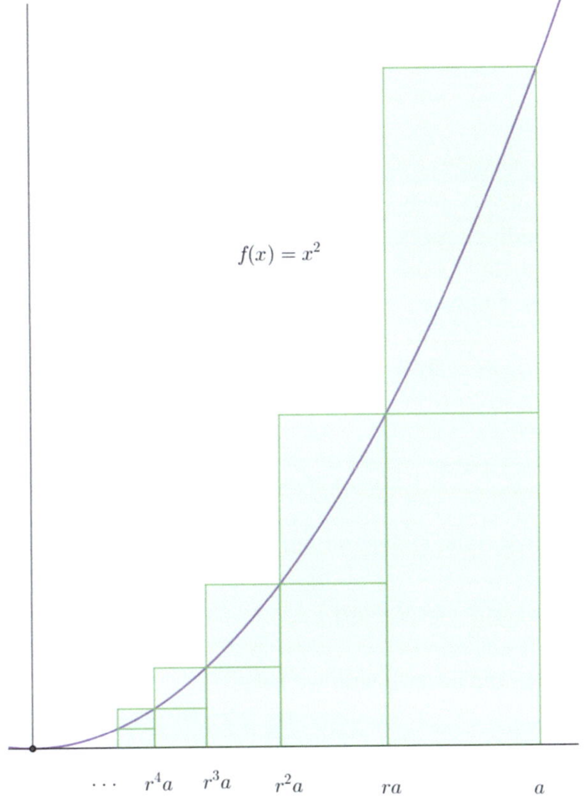

Fig. 4.5 Fermat's division of the interval for the parabola

and we can simplify the result, so that the area A under the function graph satisfies the following inequality:

$$A < \frac{a^3}{1 + r + r^2}.$$

For the lower sum, we get

$$\begin{aligned} U &= a(1-r) \cdot (ar)^2 + a\left(r - r^2\right) \cdot \left(r^2 a\right)^2 + a\left(r^2 - r^3\right) \cdot \left(r^3 a\right)^2 \\ &\quad + \cdots + a\left(r^n - r^{n+1}\right) \cdot \left(r^{n+1} a\right)^2 + \cdots \\ &= r^2 \cdot a^3(1-r) \cdot \left(1 + r^3 + r^6 + r^9 + \cdots + r^{3n} + \cdots\right) = r^2 \cdot \frac{a^3}{1 + r + r^2} \end{aligned}$$

Thus, we can write

$$r^2 \cdot \frac{a^3}{1+r+r^2} < A < \frac{a^3}{1+r+r^2}.$$

Letting r approach 1, we see that $A = a^3/3$.

The advantage of this method is that we do not have to use new sum formulas to obtain corresponding results for higher powers $y = x^k$, as was the case with the first method. The method works for all power functions with natural exponents k. For the upper sum, we get:

$$\begin{aligned} O &= a(1-r)\cdot a^k + a\left(r-r^2\right)\cdot (ra)^k + a\left(r^2-r^3\right)\cdot \left(r^2 a\right)^k + \cdots \\ &\quad + a\left(r^n - r^{n+1}\right)\cdot \left(r^n a\right)^k + \cdots \\ &= a^{k+1}(1-r)\cdot \left(1 + r^{k+1} + r^{2(k+1)} + r^{3(k+1)} + \cdots + r^{n(k+1)} + \cdots\right) = \frac{a^{k+1}(1-r)}{1-r^{k+1}} \end{aligned}$$

Here, the formula for finite geometric series

$$1 - r^{k+1} = (1-r)\left(1 + r + r^2 + r^3 + \dots + r^k\right)$$

helps us further. For the lower bound we have

$$\begin{aligned} U &= a(1-r)\cdot (ra)^k + a\left(r-r^2\right)\cdot \left(r^2 a\right)^k + a\left(r^2-r^3\right) \\ &\quad \cdot \left(r^3 a\right)^k + \cdots + a\left(r^n - r^{n+1}\right)\cdot \left(r^{n+1} a\right)^k + \cdots \\ &\quad a^{k+1}(1-r) \\ &\quad \cdot r^k\left(1 + r^{k+1} + r^{2(k+1)} + r^{3(k+1)} + \cdots + r^{n(k+1)} + \cdots\right) \\ &= \frac{r^k a^{k+1}(1-r)}{1-r^{k+1}}. \end{aligned}$$

Since the area A lies between the upper and lower bounds we have

$$r^k \cdot \frac{a^{k+1}}{1+r+r^2+\cdots+r^k} < A < \frac{a^{k+1}}{1+r+r^2+\cdots+r^k}.$$

Letting r approach 1 again, we see that $A = \frac{a^{k+1}}{k+1}$. This was achieved without the application of new sum formulas for higher and higher powers.

Remark 4.1

Fermat's second method has some clear disadvantages compared to the first, which uses equidistant interval divisions. Here, in the second method, two nested limit processes are used to calculate the area. First, for $r > 1$ an infinite geometric series is calculated to give the upper sum or the lower sum. Then the factor r has to be pushed towards 1. This is of course a conceptual extra challenge in the learning process, which does not occur with the equidistant interval divisions. On the other hand, the second method has the advantage that it can be used for significantly more functions. Thus, it has a greater explanatory potential. We also do not need to constantly prove new sum formulas to integrate power functions with ever larger exponents.

4.5 Further Development of the Second Fermat Method

It turns out that the second Fermat method can also be used for other functions that Fermat himself did not tackle at the time of developing his method. This makes the method particularly interesting for us. This is precisely the aim of this book, namely to show where historical methods still have untapped potential to explore this and possibly uncover limits.

First, we consider the integral under the hyperbola $A = \int\limits_1^a \frac{1}{x} dx$.

Fermat writes about the quadrature of hyperbolas, after he has calculated the first example (1896, p. 4):

> "The proof is the same in all other cases. Only for the primary hyperbola, i.e. the simple hyperbola or the hyperbola of Apollonius $\left(y = \frac{1}{x}\right)$, the method fails. The reason for this is that the rectangles EH, IO, MN, etc. (see Fig. 4.3) are always the same size. The terms that make up the sequence have no difference, because they are all the same. But this difference is precisely the key to the whole thing."

However, it turns out that Fermat's method can also be applied here. Here, it is important that we only define the integral over a finite interval. We set the interval end point a equal to a power of the chosen factor r, like $a = r^n$ and now consider the finite interval division, $1, r, r^2, r^3, \ldots, r^n = a$, for a $r > 1$. The upper sum for the area under the hyperbola is then the following finite sum

$$O = (r-1) \cdot 1 + \left(r^2 - r\right) \cdot \left(\frac{1}{r}\right) + \left(r^3 - r^2\right) \cdot \left(\frac{1}{r^2}\right) + \cdots$$
$$+ \left(r^{n+1} - r^n\right) \cdot \left(\frac{1}{r^{n-1}}\right) = (r-1)$$
$$\cdot (1 + 1 + \cdots + 1) = (r-1)n$$

where we have again multiplied the corresponding interval width with the largest function value in the interval before we summed up all terms (see Fig. 4.6).

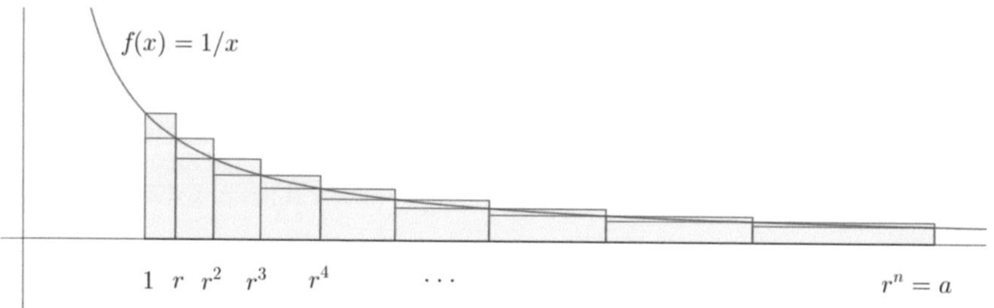

Fig. 4.6 Fermat's second integration method for the hyperbola

Correspondingly, we get for the lower sum

$$U = (r-1)\cdot\frac{1}{r} + \left(r^2 - r\right)\cdot\left(\frac{1}{r^2}\right) + \left(r^3 - r^2\right)\cdot\left(\frac{1}{r^3}\right) + \cdots$$
$$+ \left(r^{n+1} - r^n\right)\cdot\left(\frac{1}{r^n}\right) = \frac{(r-1)n}{r}$$

and we can give upper and lower bounds for the area A under the function graph of the hyperbola as follows

$$\frac{(r-1)n}{r} < A < (r-1)n.$$

Using now $r^n = a$, we get with today's terminology $n\ln(r) = \ln(a)$ and thus

$$\ln(a)\frac{r-1}{r\ln(r)} \leq A \leq \ln(a)\frac{r-1}{\ln(r)}.$$

If we can now determine the limit $C = \lim_{r\to 1}\frac{r-1}{\ln(r)}$, we have a formula for the integral under the hyperbola. In any case, we see that

$$\int_1^a \frac{1}{x}dx = C\ln(a),$$

assuming that the aforementioned limit exists. For Fermat, this limit was unknown because the Euler number e and thus the natural logarithm had not yet been explored at that time. It was not until Euler (1748, Chap. 7) that the value was determined in 1748. With the help of L'Hopital's rule, we can easily find today

$$\lim_{r\to 1}\frac{r-1}{\ln(r)} = \lim_{r\to 1}\frac{(r-1)'}{(\ln(r))'} = \lim_{r\to 1}\frac{1}{\frac{1}{r}} = \lim_{r\to 1} r = 1.$$

However, this approach assumes that we know the derivative of the logarithm, which is equivalent to knowing the hyperbolic integral due to the Fundamental Theorem of Calculus. Nevertheless, we were able to easily demonstrate the qualitative result $\int_1^a \frac{1}{x}dx = C\ln(a)$ using Fermat's second method, which is already an astonishing result in itself.

Above, when we used $n\ln(r) = \ln(a)$, we had used the logarithm as the inverse of the exponential function. If we set $h = \ln(r)$ or equivalently $e^h = r$, we can write

$$\frac{r-1}{\ln(r)} = \frac{e^h - 1}{h}.$$

Now we have

$$C = \lim_{r\to 1}\frac{r-1}{\ln(r)} = \lim_{h\to 0}\frac{e^h - 1}{h} = G = 1,$$

because h approaches zero when r approaches 1, and vice versa. The latter of the two limits G is precisely the derivative of the exponential function at $x = 0$. We had already calculated this limit in Chap. 3 on Gregory's method and also used it when integrating the exponential function earlier in this chapter.

Another corollary now presents itself with Fermat's second method, namely the integral of the logarithm itself. This time we are again interested in the interval $[0, a]$. Again, we consider the interval partition

$$\ldots, ar^n, ar^{n-1}, \ldots ar^3, ar^2, ar, a \text{ for } r < 1$$

and now study the upper and lower sums in the above sense (see Fig. 4.7). We obtain:

$$\begin{aligned}
O &= a(1-r)\cdot\ln(a) + a\left(r - r^2\right)\cdot\ln(ra) + a\left(r^2 - r^3\right)\cdot\ln\left(r^2a\right) + \cdots \\
&\quad + a\left(r^n - r^{n+1}\right)\cdot\ln\left(r^n a\right) + \cdots \\
&= a(1-r)\left(\ln(a) + r\ln(ra) + r^2\ln(r^2a) + \cdots r^n\ln\left(r^n a\right) + \cdots\right) \\
&= a(1-r)\ln(a)\left(1 + r + r^2 + \cdots r^n + \cdots\right) \\
&\quad + a(1-r)\ln(r)r\left(1 + 2r + 3r^2 + \cdots nr^{n-1} + \cdots\right) \\
&= \frac{a(1-r)\ln(a)}{1-r} + \frac{a(1-r)r\ln(r)}{(1-r)^2} = a\ln(a) + ar\frac{\ln(r)}{(1-r)}.
\end{aligned}$$

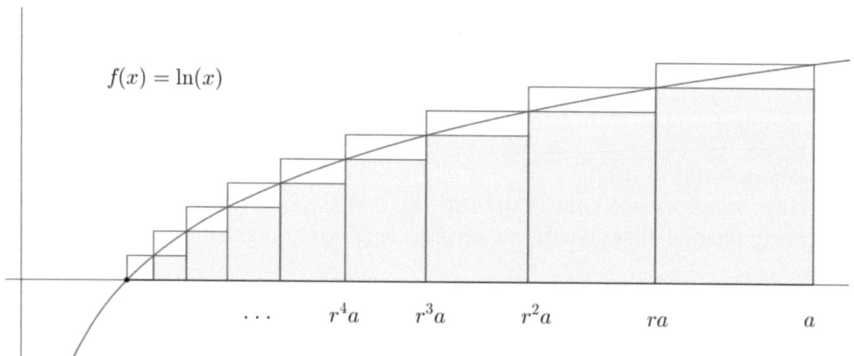

Fig. 4.7 Fermat's second integration method applied to the logarithm function

In the penultimate transformation, we used the following:

$$\begin{aligned}\frac{1}{(1-r)^2} &= \frac{1}{1-r}\cdot\frac{1}{1-r} = \left(1+r+r^2+r^3+\cdots\right)\cdot\left(1+r+r^2+r^3+\cdots\right)\\ &= 1+2\cdot r+3\cdot r^2+4\cdot r^3+\cdots nr^{n-1}+(n+1)r^n+\cdots.\end{aligned}$$

The power r^n gets the factor $n+1$, because in total $n+1$ combinations r^j and r^{n-j} from the two brackets together generate the power r^n. For the lower sum, we get a very similar result, namely

$$U = a\ln(a) + a\frac{\ln(r)}{(1-r)}.$$

However, we know that $C = \lim_{r\to 1}\frac{r-1}{\ln(r)} = 1$, and we obtained the well-known integral

$$\int_0^a \ln(x)dx = a\ln(a) - a.$$

Other functions can also be integrated using exactly the same tools. We consider $f(x) = x\ln(x)$. Here too, the method described above can be applied.

$$\begin{aligned}O &= a(1-r)\cdot a\ln(a) + a\left(r-r^2\right)\cdot ra\ln(ra) + a\left(r^2-r^3\right)\cdot r^2a\ln\left(r^2a\right)+\cdots\\ &\quad + a\left(r^n-r^{n+1}\right)\cdot r^na\ln\left(r^na\right)+\cdots\\ &= a^2(1-r)\left(\ln(a)+r^2\ln(ra)+r^4\ln\left(r^2a\right)+\cdots r^{2n}\ln\left(r^na\right)+\cdots\right).\end{aligned}$$

Now we use the rule $\ln(ab) = \ln(a) + \ln(b)$ and obtain

$$\begin{aligned}
\mathrm{O} &= a^2(1-r)\ln(a)\left(1 + r^2 + r^4 + \cdots r^{2n} + \cdots\right) \\
&\quad + a^2(1-r)\ln(r)r^2\left(1 + 2r^2 + 3r^4 + \cdots + (n+1)r^{2n} + \cdots\right) \\
&= \frac{a^2(1-r)\ln(a)}{1 \quad r^2} + \frac{a^2(1-r)r^2\ln(r)}{\left(1-r^2\right)\left(1-r^2\right)} = \frac{a^2\ln(a)}{1+r} + \frac{a^2r^2}{(1+r)^2}\frac{\ln(r)}{1-r}.
\end{aligned}$$

Here the lower sum is then (see Fig. 4.8)

$$U = \frac{a^2r\ln(a)}{1+r} + \frac{a^2r}{(1+r)^2}\frac{\ln(r)}{1-r}.$$

But since we know the limit $C = \lim_{r\to1}\frac{r-1}{\ln(r)} = 1$, we can conclude that

$$\int_0^a x\ln(x)dx = \frac{a^2\ln(a)}{2} - \frac{a^2}{4}.$$

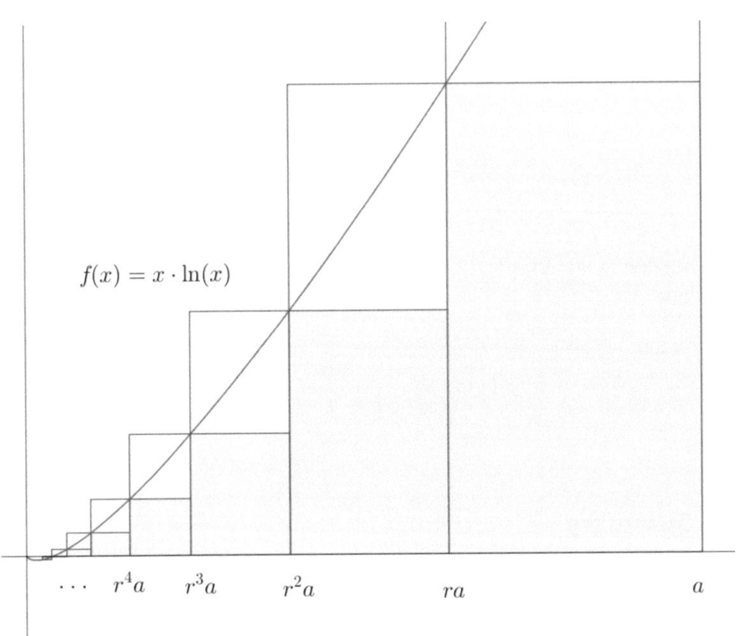

Fig. 4.8 Fermat's second integration method applied to the function $f(x) = x \cdot \ln(x)$

We can proceed in a similar way with the function $f(x) = x^k \ln(x)$ for natural numbers k. Again, we can specify the upper and lower sums and then calculate the limit. For the upper sum, we have:

$$\begin{aligned} O &= a(1-r)\cdot a^k \ln(a) + a\left(r-r^2\right)\cdot (ra)^k \ln(ra) + a\left(r^2-r^3\right)\cdot \left(r^2 a\right)^k \ln\left(r^2 a\right) \\ &\quad + \cdots + a\left(r^n - r^{n+1}\right)\cdot \left(r^n a\right)^k \ln\left(r^n a\right) + \cdots \\ &= a^{k+1}(1-r)\Big(\ln(a) + r^{k+1}\ln(ra) + r^{2k+2}\ln\left(r^2 a\right) \\ &\quad + \cdots r^{kn+n}\ln\left(r^n a\right) + \cdots). \end{aligned}$$

Again, we use the rule $\ln(ab) = \ln(a) + \ln(b)$ and obtain

$$\begin{aligned} \mathrm{O} &= a^{k+1}(1-r)\ln(a)\Big(1 + r^{k+1} + r^{2(k+1)} + \cdots r^{n(k+1)} + \cdots\Big) \\ &\quad + a^{k+1}(1-r)\ln(r)r^{k+1}\Big(1 + 2r^{k+1} + 3r^{2(k+1)} + \cdots + (n+1)r^{n(k+1)} + \cdots\Big) \\ &= \frac{a^{k+1}(1-r)\ln(a)}{1-r^{k+1}} + \frac{a^{k+1}(1-r)r^{k+1}\ln(r)}{\left(1-r^{k+1}\right)\left(1-r^{k+1}\right)} \\ &= \frac{a^{k+1}\ln(a)}{1+r+r^2+\cdots+r^k} + \frac{a^{k+1}r^{k+1}}{\left(1+r+r^2+\cdots+r^k\right)^2}\frac{\ln(r)}{1-r}. \end{aligned}$$

Accordingly, the lower sum is calculated as

$$U = \frac{a^{k+1}r^k\ln(a)}{1+r+r^2+\cdots+r^k} + \frac{a^{k+1}r^k}{\left(1+r+r^2+\cdots+r^k\right)^2}\frac{\ln(r)}{1-r}.$$

By now, however, we know the limit $C = \lim_{r\to 1}\frac{r-1}{\ln(r)} = 1$, and thus we can conclude that

$$\int_0^a x^k \ln(x)dx = \frac{a^{k+1}\ln(a)}{k+1} - \frac{a^{k+1}}{(k+1)^2}.$$

4.6 Summary

In this chapter, we have learned about two methods of integration developed by Fermat. The first method is still found in textbooks in both schools and colleges and universities, although not consistently. The second method is only found in historical texts, but it far surpasses the first in terms of its scope of application but also brings with it greater conceptual challenges. At the same time, it still contains potential for new applications that were not noted by Fermat. In our opinion, this potential is well suited for teacher training, as it also brings together complex ideas. Infinite sums, combinatorial arguments, and limit considerations are incorporated into the line of argumentation. Further reasons speak for the use of the topic in teacher training: On the one hand, future teachers can learn here that there are several approaches to integration, and on the other hand, the methods presented here are suitable for sparking a discussion about the advantages of the individual methods (equidistant and non-equidistant interval divisions), both purely mathematically and didactically. By using different methods of interval division, we have the opportunity to point out the fact of the independence of the interval division in the Riemann integral, which remains difficult to explain if we only use a single division method. Also, interested students can work with the material presented here when it comes to a deeper understanding of the concept of integration and especially when the idea of integration as "antidifferentiation" is to be expanded.

Bibliography

Burn, R. P. (1999): *Integration, a genetic introduction,* Nordisk matematikkdidaktikk, Nr. (7)1.

Brandt, D., Freudigmann, H., Greulich, D., Riemer W.: Lambacher Schweizer – Mathematik für Gymnasien, Leistungskurs Rheinlandpfalz. Ernst Klett Verlag Stuttgart, Leipzig (2011)

Euler, L. (1748), Introductio in analysin infinitorum. Vol 1., English translation 1988. Berlin: Springer

Fermat P. (1896): Sur la transformation et la simplification des équations de lieux, pour la comparaison sous toutes les formes des aires curvilignes, soit entres elles, soit avec des rectilignes, et en même temps sur l'emploi de la progression géométrique pour la quadrature des paraboles et hyperboles à l'infini. In: Tannery, P., Henry C. Œuvres de Fermat, Tome troisième, Gauthier-Villars et fils, 216–237. English translation: http://science.larouchepac.com/fermat/

Kirfel, C. (2017A): FERMATS Methoden zur Integration, Der Mathematikunterricht, Friedrich Verlag (2017/4), pp. 44–50.

Körner, H., Lergenmüller A., Schmidt, G.: Mathematik Neue Wege—Arbeitsbuch für Gymnasien, Analysis II. Schroedel, Braunschweig (2011)

Riemann B., Weber H., & Dedekind, R. (1876): *Bernhard Riemann's Gesammelte mathematische Werke und Wissenschaftlicher Nachlass. Leipzig*:

Roth J., Siller H.S., Bestand und Änderung, Grundvorstellungen entwickeln und nutzen, Mathematik lehren, Friedrich Verlag, 199/2016

The Resection Method of Leibniz and Its Inversion

5

5.1 Introduction

In his 1676 manuscript "De quadratura arithmetica circuli ellipseos et hyperbolae," Leibniz (2016) describes the Resection Method for determining areas under curves. Although the method also uses tangents and thus the derivative, it is not identical to the fundamental theorem of calculus, but can be derived from it as a special case. With this method, Leibniz establishes a connection between the area under given curves and other curves, which are often easier to determine. This allows him to determine areas under many known curves. After introducing the Resection Method, we turn to the so-called Fringe Method. This also establishes a connection between areas under different curves, and it turns out that it inverts the Resection Method. This is the main result of this chapter. Particular emphasis is placed on the geometric interpretation of the Fringe Method, which also sheds new light on the Resection Method. A series of examples serves to illustrate the method.

In the 16th and 17th centuries, integral calculus was developed by great minds such as Descartes, Fermat, Cavalieri, Barrow, Wallis, Roberville, Newton, and Leibniz. Many of the ideas emerged simultaneously, in response to pressing questions of contemporary natural sciences. Similar ideas and approaches to integral calculus were being worked on in many parts of Europe. In the modern presentation of this subject in schools and introductory university courses, the ideas leading to the fundamental theorem of calculus predominate, while other approaches are hardly noticed and thus become the subject of historical interest. Leibniz's Resection Method is such an independent approach to solving the problem of calculating areas between curves. In his 1676 paper, Leibniz (2016) extensively develops his method and determines areas under many of the curves known at the time, including a circular arc, where he comes across the so-called Gregory-Leibniz series for the circle number π. Power functions, logarithmic functions, cycloids, and conic sections can also be approached with the Resection Method.

© The Author(s), under exclusive license to Springer-Verlag GmbH, DE, part of Springer Nature 2026

C. Kirfel, *Side Paths in the History of Mathematics*, Mathematics Study Resources 21, https://doi.org/10.1007/978-3-662-72918-2_5

Roberval (1602–1675) had already developed a very similar Resection Method in 1646, so Leibniz was not the first to use such a method. However, it is very likely (Probst 2016) that Leibniz developed his method independently. Barrow had also developed a transmutation theorem before Leibniz, but it is also assumed here that Leibniz discovered his method independently (Probst 2017). In the following, we use the Leibnizian representation and not that of Roberval or Barrow, because Leibniz used and promoted this method so extensively.

5.2 Leibniz's Resection Method

In 1676, Leibniz wrote a comprehensive manuscript that was not published until much later (Leibniz 2016). At this point, we would like to introduce the important Theorem 7 from the Leibnizian paper, which describes the Resection Method. Leibniz himself calls this theorem the transmutation theorem, or the transformation theorem. Here is the corresponding excerpt from Leibniz's paper (ibid., p. 33, the author's translation):

Theorem 7
"If from any point of a certain curve we draw on one side of a right angle located in the same plane, perpendicular ordinates, on the other side tangents, and from the points of intersection of the tangents with their possibly extended ordinates perpendicular lines are drawn, and another curve passes through the intersections of these perpendicular lines and ordinates, the area enclosed by the axis (to which the ordinates are drawn), the two outermost ordinates, and the second curve will be twice as large as the area enclosed by the first curve and the two lines that connect the outermost points of it with the center of the given right angle." (see Fig. 5.1).

In modern language we could say: "A curve is given in a coordinate system. From any point on the curve parallels to the x-axis are drawn. In addition, tangents to the curve are drawn and these tangents are intersected with the x-axis. The distance between the origin and the intersection points of the tangents and the x-axis are now translated to the parallels mentioned above and these lengths are marked off on the parallels starting from the y-axis. These intersections form a new curve. The area enclosed by the y-axis, the two uttermost parallels and the second curve will be twice as large as the area enclosed by the first curve and the two lines that connect the outermost points of it with the origin." (see Fig. 5.1).

Afterwards, Leibniz writes a few lines about his own assessment of his result:

> "As for the theorem itself, I believe it is one of the most general and useful in geometry. For it is so comprehensive that it fits all curves, even if they are random or drawn at will without a specific law, and that, given any figure, it shows infinitely many others whose area determination depends on the first and vice versa. But it can also be counted among the most fruitful theorems of geometry, for from here immediately the quadratures of all paraboloids and hyperboloids to infinity are proven, or of figures in which the ordinates or their powers

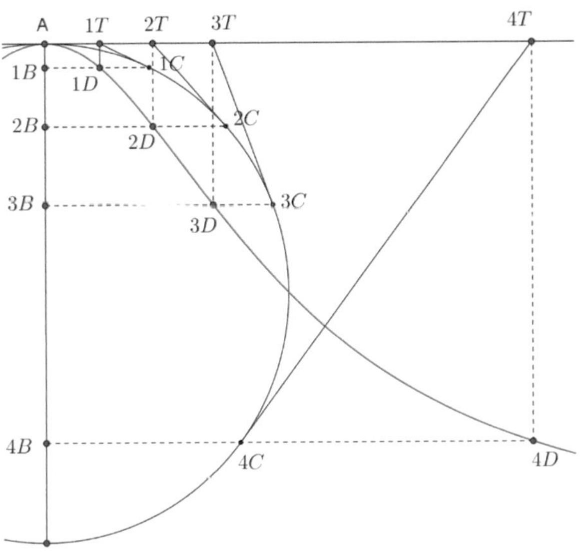

Fig. 5.1 Sketch from Leibniz's manuscript

> are in a multiplicative or submultiplicative direct or reciprocal ratio to the abscissas or the powers of the abscissas. And we have, to leave aside infinitely many other solved or hypothetical quadratures, transformed the circle in any case and any conic section with a center with its help into a rational figure, and from here we derived the arithmetic quadrature of the entire circle and of any part, as well as the true and perfect analytical expression of a circular arc from a given tangent; this treatise deals with the proof of these things... from which a vast field of new discoveries opens up, the foundations of which we provide here …"

Despite the enthusiasm that Leibniz expresses here about his work, it does not get published. It appears that the discovery of the fundamental theorem of integral and differential calculus also falls precisely into this period. This fundamental theorem solves the problem of integration much more generally than the Method of Resection and far surpasses Leibniz's achievements. Therefore, this topic is a particularly suitable subject for the present book project.

Instead of Leibniz's proof, we would like to present a modern proof.

5.3 Modern Proof

Compared to the Leibnizian representation in Fig. 5.1, the illustration in Fig. 5.2 is rotated by 90°. This makes the calculations a bit simpler for us. We want to determine the area C of the segment OAD, which is bounded by the curve AD, formed by the graph of the function $f(x)$, and the lines DO and OA. The integral under the curve between a and b can then be determined from the area C and the

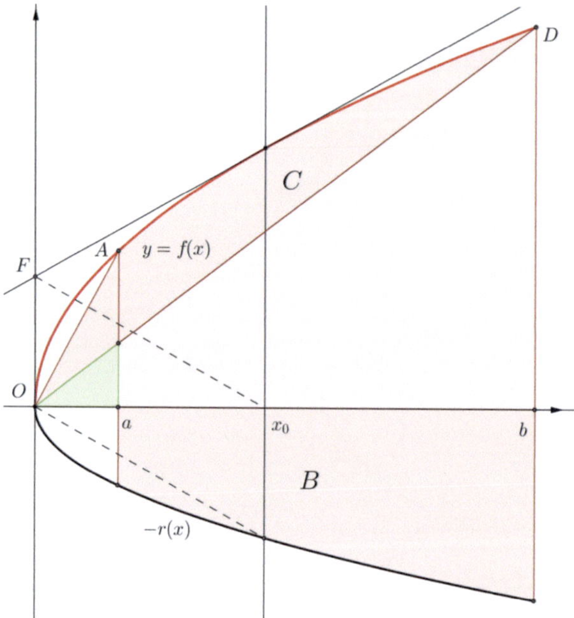

Fig. 5.2 Modern proof of the transmutation theorem (Resection Method)

area of various triangles. Following the description in the transmutation theorem, we start with a point $(x_0, f(x_0))$ on the graph of the function $y = f(x)$ and erect the tangent to the function graph there. The coordinate axes form the right angle that occurs in the Leibnizian transmutation theorem. We intersect the tangent with the y-axis and plot the "tangent segment" OF below the x-axis and beneath the point on the curve, so that the two areas C and B do not overlap.

The equation of the tangent at the point $(x_0, f(x_0))$ is

$$y - f(x_0) = f'(x_0)(x - x_0).$$

Therefore, $|FO| = f(x_0) - x_0 f'(x_0)$. Here we have used $x = 0$ to calculate the y-value of the intersection point F of the tangent with the y-axis. This distance $|FO|$ is now plotted below the x-axis at x_0, which corresponds to a parallel translation of the segment OF as indicated by the dashed lines. If this process is carried out for every point of the function, then a new curve is created below the x-axis. Leibniz calls the new function $r(x) = f(x) - xf'(x)$ the resection. In Fig. 5.2, we have plotted the resection downwards so that the areas under consideration do not overlap. Leibniz himself plots the resection on the same side as the original function graph, i.e., above the x-axis. The areas are of course equal. The function graph in Fig. 5.2 actually represents $-r(x)$. However, we are only interested in the area, which is the same in both cases.

If we want to calculate the area of the segment C under the original curve with the boundary lines OA and OD, we get

$$C = \int_a^b f(x)dx - \frac{bf(b)}{2} + \frac{af(a)}{2}.$$

Here, the triangle with the base b and the height $f(b)$ was subtracted and the triangle with base a and height $f(a)$ was added. The green overlapping triangle is therefore added and subtracted again, so it does not contribute. The area under the resection, the new function below the x-axis, over the interval $[a, b]$ is then calculated as

$$B = \int_a^b \big(f(x) - xf'(x)\big)dx.$$

By means of integration by parts, we have:

$$\begin{aligned} B &= \int_a^b \big(f(x) - xf'(x)\big)dx = \int_a^b f(x)dx - \int_a^b xf'(x)dx \\ &= 2\int_a^b f(x)dx - [xf(x)]_a^b = 2\left(\int_a^b f(x)dx - \frac{bf(b)}{2} + \frac{af(a)}{2}\right) = 2C. \end{aligned}$$

With this, we have proven Leibniz's result using modern methods, namely integration by parts (see Ullrich 2017). Our current rule for integration by parts is based on the fundamental theorem of calculus, which Leibniz did not have at his disposal. The use of integration by parts at this point may therefore seem ahistorical or anachronistic, but on the other hand, it is very efficient. In the next section we will see how Leibniz applies his method to the area of a circle.

5.4 Leibniz's Calculation of π

Leibniz developed his method in order to calculate π. Already in the illustration of Theorem 7 (Fig. 5.1), Leibniz uses a circular arc, even though he wants to describe a general curve. This suggests that one of his higher goals was precisely the determination of the area under the circular arc, or in other words, the determination of π. As announced above, he also determines many areas under other curves. We now show how to determine the area of the circular arc using the Resection Method.

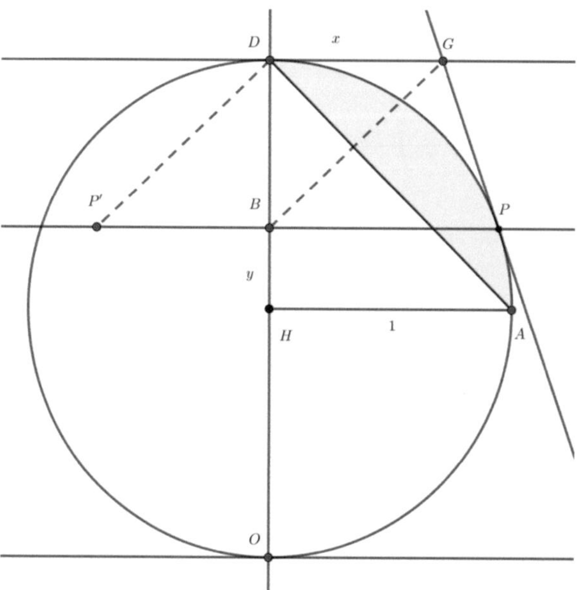

Fig. 5.3 The Method of Resection on the circle

First, he considers a quarter circle HAD (see Fig. 5.3) with radius 1 and places a tangent at the circle point P, which he intersects with the line $y = 2$ at G (see Figs. 5.3 and 5.4). The line $y = 2$ is also a tangent to the circle. $DG = x$ is then the segment that is central to the Resection Method. Because G lies on both tangents, the angles $\angle DHG$ and $\angle GHP$ are equal and we can write $\angle DHG = \angle GHP = \alpha$ (see Fig. 5.4). Then it follows that

$$2\alpha + \delta = 180°,$$

where δ represents the angle $\angle PHO$. Since the triangle $PH0$ is isosceles we have also

$$2\gamma + \delta = 180°,$$

so all in all $\alpha = \gamma$. This can also be seen as a consequence of the result on peripheral and central angles in a circle. Thus, the right-angled triangles OBP and DHG are similar and we have

$$\frac{x}{1} = \frac{BP}{OB} = \frac{\sqrt{y(2-y)}}{y} = \sqrt{\frac{2-y}{y}},$$

because the theorem on altitudes in a right triangle gives us $BP^2 = y(2-y)$, a result we are going to explain below. Using $y = OB$ we get after some

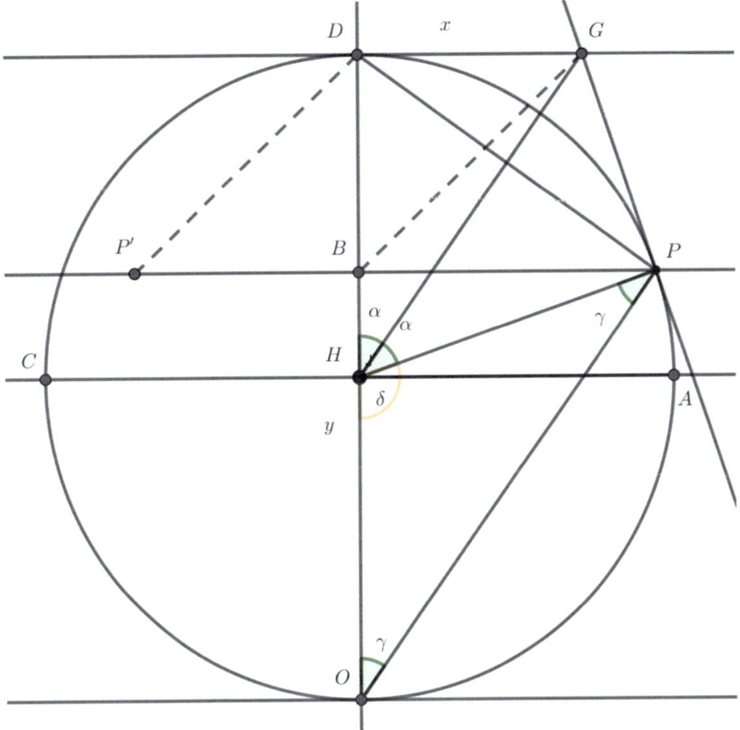

Fig. 5.4 Calculation of π, part 1

transformations

$$x^2y = 2 - y \text{ or } y\left(1 + x^2\right) = 2 \text{ i.e. } y = \frac{2}{1 + x^2}.$$

If you plot x at the height of P to the left of the y-axis (see Fig. 5.5), and if you do this for every point P of the quarter circle DA, you get the curve described by Leibniz, the resection, which encloses the area B with the y-axis and the boundary line $y = 1$. Because the area B in Fig. 5.5 lies above the line $y = 1$, we calculate its area as the following integral

$$B = \int_{-1}^{0} \left(\frac{2}{1 + x^2} - 1 \right) dx = 2 \int_{0}^{1} \frac{dx}{1 + x^2} - 1.$$

Now, as a result of the Resection Method, B has twice the area of the circular sector C. This sector, together with the triangle $T = DHA$, whose area is ½, forms

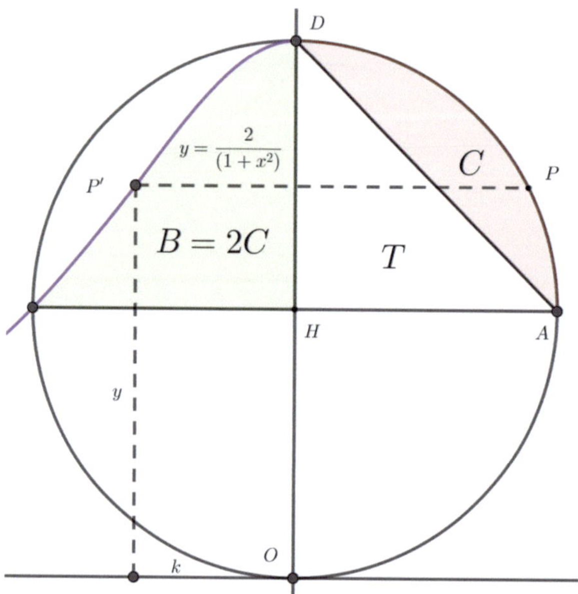

Fig. 5.5 Calculation of π, part 2

exactly a quarter circle and we have

$$\frac{\pi}{4} = T + C = \frac{1}{2} + \frac{B}{2} = \int_0^1 \frac{dx}{1+x^2}.$$

Leibniz now treats the integrand $\frac{1}{1+x^2} = 1 - x^2 + x^4 - x^6 \ldots$ as a geometric series and "integrates" it term by term (Leibniz 2016, pp.135–137 and 161), referring to Roberval, Fermat, and Wallis. (In the next section we will see how Leibniz himself, using the Resection Method, achieves the required result of integrating monomials x^n.) Here we obtain the well-known series

$$\frac{\pi}{4} = \int_0^1 (1 - x^2 + x^4 - x^6 + \cdots)dx = 1 - \frac{1}{3} + \frac{1}{5} - \frac{1}{7} + \ldots$$

To understand the relationship $BP^2 = y(2-y)$, we have cited the so-called altitude theorem (see Fig. 5.6). It is the following result on the semicircle. The triangles AED and EBD are similar. Therefore, $\frac{h}{a} = \frac{b}{h}$ or $h^2 = ab$, which in Fig. 5.4 then means $BP^2 = y(2-y)$.

Leibniz examines the series $1 - \frac{1}{3} + \frac{1}{5} - \frac{1}{7} + \cdots$ in detail and provides several theorems about this series and the constant π. To integrate the series $1 - x^2 + x^4 -$

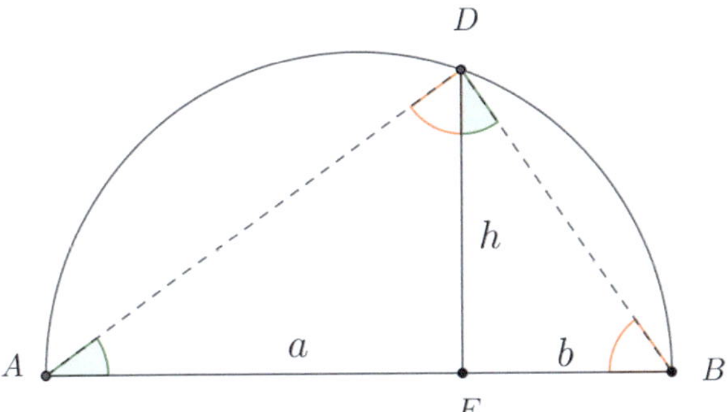

Fig. 5.6 The altitude in a right triangle

$x^6 + \cdots$, we must know how to integrate a monomial, i.e., a power function of the form x^n. We will explain this in the following section.

5.5 Power Functions Using the Method of Resection

Leibniz was not satisfied with just calculating the circle number π. We would like to show here that his method has a much wider range and that numerous other types of functions can be integrated using the Method of Resection. First, we consider—as promised above—power functions $y = x^n$ for $n > 1$. Here, the slope of the tangent at the point $(x, f(x))$ is equal to $f'(x) = nx^{n-1}$. In Fig. 5.7, we see that the tangent section T under the tangent from the intersection point G of the tangent with the x-axis to the base point $W = (x, 0)$ under the curve point $P = (x, f(x))$ is $T = \frac{f(x)}{f'(x)}$. Thus, $T = \frac{x^n}{nx^{n-1}} = \frac{x}{n}$. The complementary tangent segment OG, which is of interest for the Method of Resection, has the length $OG = x - \frac{x}{n} = \frac{n-1}{n}x$. This distance OG is now translated to the left of the y-axis at the height of the curve point P, see Fig. 5.7. Leibniz draws it on the same side of the y-axis as the function graph. To avoid overlapping areas, we have translated the segment to the other side of the y-axis here.

The distance of the original curve point P from the y-axis is x, while the distance of the new point (now to the left of the y-axis) is only $\frac{n-1}{n}x$. The value is thus multiplied by the factor $\frac{n-1}{n}$. According to Cavalieri's principle, the area is $B = \frac{n-1}{n}(D + C)$, where D represents the area of the triangle OPQ.

Due to the Method of Resections, we have $B = 2C$. Thus, we get

$$2C = \frac{n-1}{n}(D + C)$$
$$(n + 1)C = (n - 1)D$$

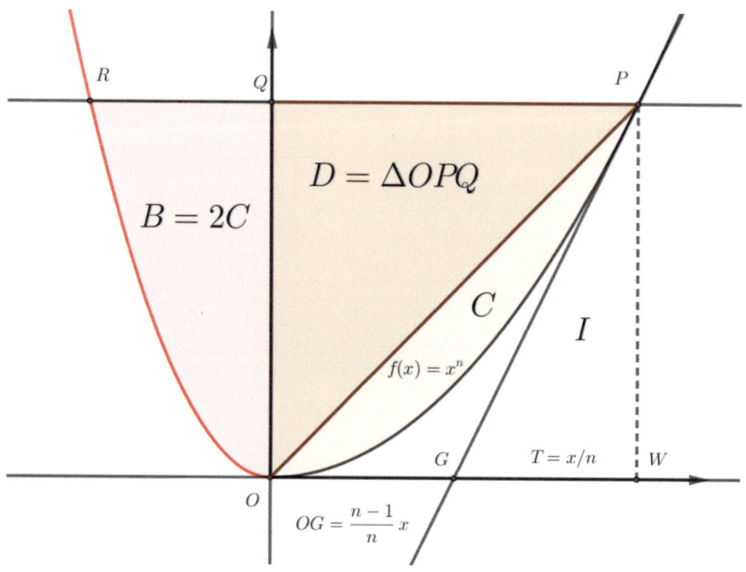

Fig. 5.7 The resection of the power function

Now, $2D = x \cdot x^n = x^{n+1}$ is the area of the rectangle *OWPQ* and the area under the curve f is

$$I = D - C = \left(1 - \frac{n-1}{n+1}\right)D = \frac{2}{n+1} \cdot \frac{x^{n+1}}{2} = \frac{x^{n+1}}{n+1},$$

as we know from analysis.

For $n = 1$ we get a straight line through the origin. Here, then, $B = C = 0$ and the integral corresponds exactly to the triangle under the straight line.

When considering the function $f(x) = x^n$, Leibniz treats the different cases, depending on whether the exponent $n > 0, n < 0, n \in \mathbb{N}$ or $n \in \mathbb{Q}$. We will do the same here, but we will not go into depth as much as Leibniz does.

In the case where $0 < n < 1$, the illustration Fig. 5.7 does not apply. Figure 5.8 shows us a typical situation. Here again $B = 2C$ and $B = \frac{n-1}{n}\left(C + \frac{a \cdot a^n}{2}\right)$, so just like above $2nC = (n-1)C + (n-1)\frac{a^{n+1}}{2}$ or $C = \frac{(n-1)a^{n+1}}{2(n+1)}$. For the integral we then have again

$$I = \int_0^a x^n dx = C + \frac{a^{n+1}}{2} = \frac{a^{n+1}}{(n+1)}.$$

For negative n neither of the illustrations Figs. 5.7 or 5.8 is correct. Therefore, we need to reconsider this case. When n is negative, we do not have a horizontal

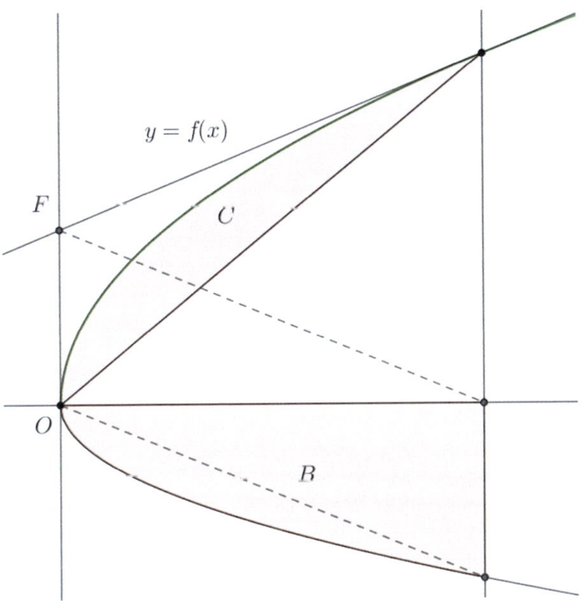

Fig. 5.8 The case $0 < n < 1$

tangent at the origin. Instead, we start with the point $A = (1, 1)$, which lies on all these power functions. We keep the coordinate grid as the right angle that occurs in the Resection Method and it follows that the area B to the left of the y-axis is twice as large as the area C to the right of the y-axis (see Fig. 5.9). These are partially curvilinearly bounded areas *OAD* and *MD'A'L*.

Also for power functions with negative exponents $n \neq -1$, Cavalieri's principle applies and we have as above $B = \frac{n-1}{n} F$ and according to the Resection Method $B = 2C$, so $(n-1)F = 2nC$ (see Figs. 5.9 and 5.10).

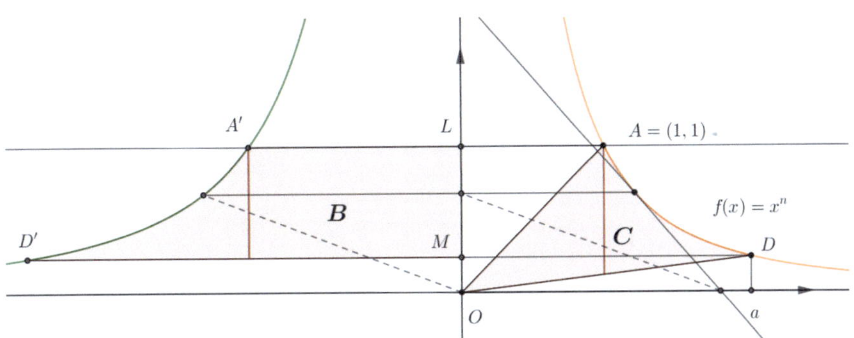

Fig. 5.9 The case of negative powers, part 1

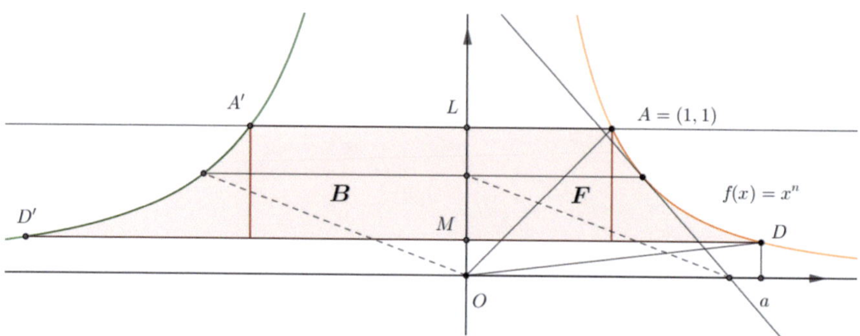

Fig. 5.10 The case of negative powers, part 2

Now $C + \frac{1}{2} + a \cdot \frac{a^n}{2} = F + a \cdot a^n$, so $C = F + \frac{a^{n+1}}{2} - \frac{1}{2}$ and thus $(n-1)F = nB = 2nC = 2n\left(F + \frac{a^{n+1}}{2} - \frac{1}{2}\right)$, so

$$F = \frac{n}{n+1}\left(1 - a^{n+1}\right).$$

Here it is crucial that $n \neq -1$. Our goal is to determine the area $I = \int_1^a x^n dx$ under the function graph over the interval $[1, a]$. In Figs. 5.10 and 5.11, it can be seen that $I + 1 = F + a^{n+1}$ and thus we have

$$I = \int_1^a x^n dx = F + a^{n+1} - 1 = \frac{n\left(1 - a^{n+1}\right)}{n+1} + a^{n+1} - 1 = \frac{a^{n+1} - 1}{n+1},$$

as we know it from analysis.

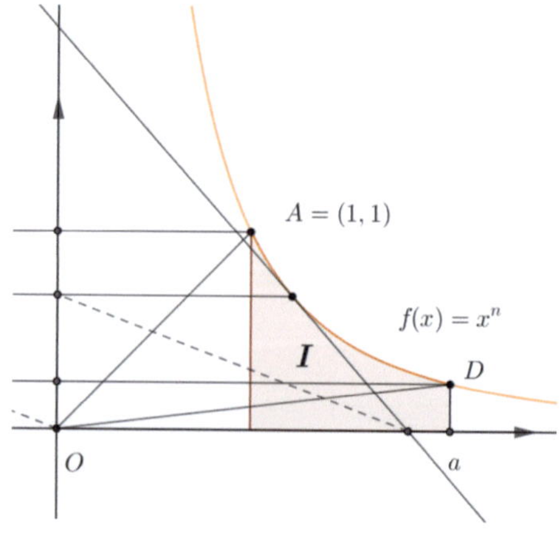

Fig. 5.11 The case of negative powers, part 3

In the case of $n = -1$ the relationship $2n\left(F + \frac{a^{n+1}}{2} - \frac{1}{2}\right) = (n-1)F$ unfortunately does not provide any further information, because the coefficients of F on both sides of the equation are the same. For the hyperbola ($n = -1$) we can therefore not draw any direct conclusions from the method of Leibniz as it is presented here.

5.6 The Exponential Function

The exponential function $y = e^x$ was not known in the time of Leibniz. It goes back to Leonhard Euler. However, Leibniz considered the logarithm function and determined its integral. Since the logarithm function is the inverse function of the exponential function, he had essentially also solved the area determination under the exponential function.

The coordinate axes again represent the right angle that occurs in the transmutation theorem. The arguments are mainly geometric in nature again. However, we need to make some preparations. We now want to determine the integral under the exponential function (the red function in Fig. 5.12) over the interval $[0, a]$, where $a < 1$. At a point $P = (x_0, e^{x_0})$ on the function graph, we draw the tangent and intersect it at the point E with the x-axis. Now we translate the line segment OE at the height of P to the left of the y-axis and find the point D. If this is done for all conceivable points P on the function graph, a new curve (orange), to the left of the y-axis, the resection curve, appears (see Fig. 5.12). The area $FKHG$ under the orange curve we denote as B, while the area $KHQP$ is named A. The tangent at the point P has the equation

$$y - e^{x_0} = e^{x_0}(x - x_0)$$

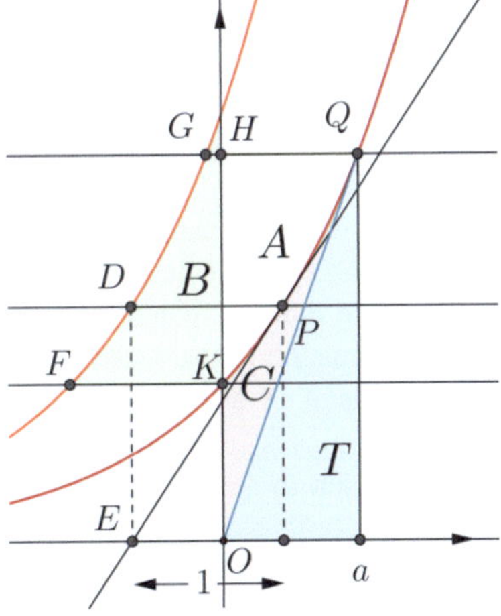

Fig. 5.12 The resection of the graph of the exponential function

For the intersection of the tangent with the x-axis, it follows that

$$x = x_0 - 1.$$

The distance from E to the origin is therefore $1 - x_0$. We have assumed $1 > a \geq x_0$. The distance from E to the base point under P is therefore equal to the constant 1. As mentioned above, Leibniz studies the logarithm function and not the exponential function. For the logarithm function we have: The tangent at a curve point $(a, \ln(a))$ intersects the y-axis at the point $(0, \ln(a) - 1)$. The difference of the y-values of the curve point and the intersection point is therefore the constant 1. Leibniz celebrates this observation as a special result.

We can therefore calculate the area *FKQG*, which reminds us of a waving flag, like a rectangle (width times height) according to Cavallieri's principle and we get

$$A + B = e^a - 1.$$

Here, $e^a - 1$ is the difference of the y-values of Q and K, the height of the "rectangle", while the width is equal to 1. Leibniz's Resection Method now gives us a relationship between the area B to the left of the y-axis and the area C, namely $B = 2C$. On the other hand we have $C + A = T$ and $C + T = I$, so $2C + A + T = I + T$ or

$$I = 2C + A = B + A = e^a - 1$$

and we have found the integral of the exponential function. Here T is the triangle $OQ(0, a)$. For $a > 1$ the line OQ will intersect the curve $y = e^x$ and thus the situation becomes somewhat more complicated. However, this can be fixed. In the case $1 < a < 2$ we can move the right angle from Leibniz's transmutation theorem to the point $(1, 0)$. The picture then looks very much like Fig. 5.12 and we find $\int_1^a e^x dx = e^a - e$. In the same manner we can find $\int_k^a e^x dx = e^a - e^k$ for $k < a < k + 1$. Adding these integrals gives us the final result $\int_0^a e^x dx = e^a - 1$.

5.7 The Cycloid

To emphasize the striking power of Leibniz's Method of Resection, we would like to present an example here that does not come from the school context. The cycloid is the curve that is created when we follow a point $P(x, y)$ on the periphery of a circle (radius $ZE = 1$), which rolls on a horizontal line (x-axis). It thus describes the movement of a bicycle valve when the wheel rolls on a straight line on the ground.

If we want to describe the instantaneous movement of the point P, we can say that at the moment it is a rotation around the contact point F with the line the wheel is rolling on. The direction of movement is therefore perpendicular to the line FP. Thales' theorem then teaches us that the tangent to the cycloid and the

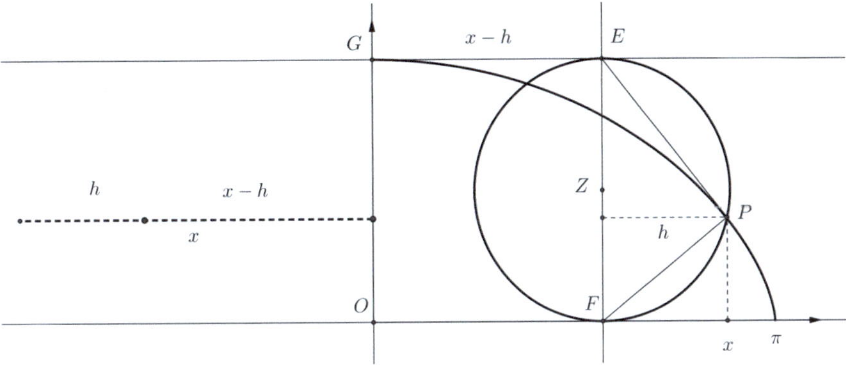

Fig. 5.13 The resection of the cycloid

line $y = 2$ intersects at the point E, which itself lies on the rolling circle through P (see Fig. 5.13).

Then we see that $GE = x - h$, where h is the height in the circle through P.

As P traverses half of the cycloid, the aforementioned height sweeps out exactly a semicircle so we get that the area under the Resection curve B satisfies $B + \frac{\pi}{2} = T + C$, where T is the area of the triangle ODG and C is the cigar shaped area in Fig. 5.14. Since $T = \pi$ we now get

$$B = T + C - \frac{\pi}{2} = C + \frac{\pi}{2}.$$

From the Resection Method we get

$$B = 2C.$$

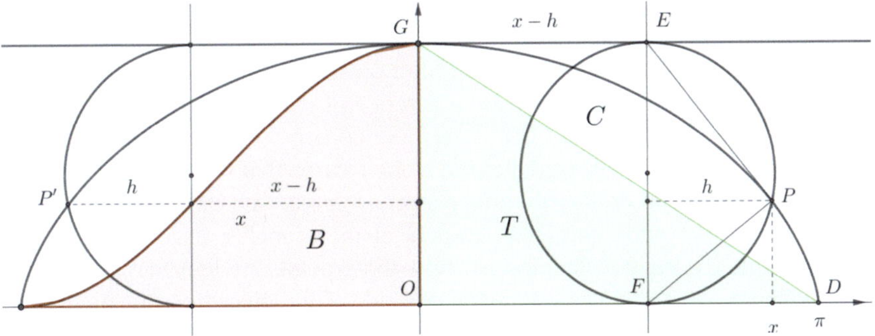

Fig. 5.14 The area under the cycloid

It follows that $C = \frac{\pi}{2}$. For the integral, the area under half of the cycloid, which we are interested in, we then have

$$I = T + C = \pi + \frac{\pi}{2} = \frac{3\pi}{2},$$

as we know from analysis.

5.8 The Fringe Method for Determining Areas Under Curves

The Fringe Method, which we would like to introduce now, is also a method that allows us to determine areas under a function graph (see Kirfel 2023 and Kirfel 2024). It turns out that it is precisely the inverse of the Resection Method. We will show this later. Now we want to describe the Fringe Method first.

With the help of the Fringe Method, we can find another curve with the same area for a given area under a function graph over an interval. In Fig. 5.15 we see an example, the graph of the function $f(x) = 1-x^2$ over the interval $[0, b]=[0, 1]$. We start with an interval division $0, \Delta, 2\Delta, \ldots, n\Delta = b$ of the interval and construct the associated Riemann rectangles over the subintervals under the function graph. These rectangles are now replaced by triangles, each having the same base as the rectangles. These bases correspond to the subintervals. The height is twice as large as that of the corresponding rectangle, with the sharp corner of the triangle pointing downwards (see Fig. 5.15). Accordingly, the area of the triangles and the corresponding rectangles is the same. The "left" side of the triangles is always chosen vertically (see Fig. 5.15).

The entirety of these triangles then has the same area as the entirety of the rectangles under the function graph. Now we move the sharp corners of the triangles on horizontal lines one after the other to the left, until each corner meets the preceding triangle (see Fig. 5.16). This movement (shearing) does not change the area of the triangles and a new connected area with the same area as the Riemann rectangles has been created. We had assumed that the function f was monotonically decreasing. Thus, the triangles become smaller for increasing x values and we push smaller triangles towards larger ones. The reverse case may possibly cause difficulties, so we disregard it and limit ourselves to monotonically decreasing functions. The new area consists of a series of triangles, with the corners of the bases always on the x-axis. The remaining corners, which originate from the original sharp corners of the triangles, lie along a curve that we want to determine. This curve is then the promised correspondence curve with the same area as the original curve. We name the sharp corners of the triangles, into which we had exchanged the Riemann rectangles, B_i. Now these points are moved on horizontal lines to the left until they come to lie on one side of the preceding triangle. In this position, we then call the corners A_i. In detail, it looks like this:

Let $B_0 = (0, -2f(0))$ be the sharp corner of the first triangle. Then the corner $B_0 = A_0$ stays where it was. The sharp corner B_1 of the second triangle is now

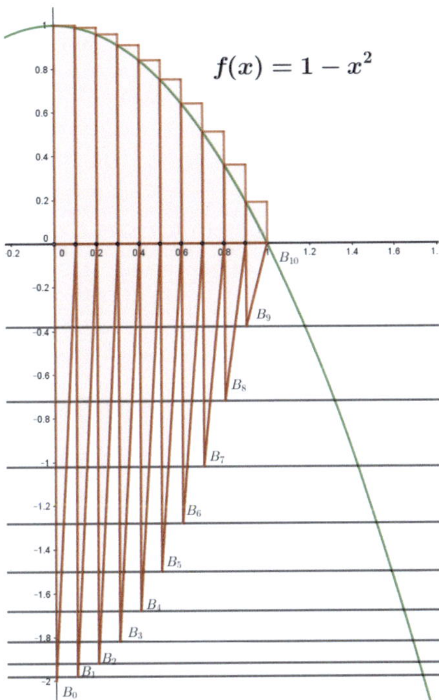

Fig. 5.15 Transforming rectangles into triangles

moved until it comes to lie on the line from B_0 to $(\Delta, 0)$. In this position, we now call the point A_1. The point B_2 is now moved horizontally until it hits the line from A_1 to $(2 \cdot \Delta, 0)$. Once there, we now call the point A_2 etc. In general, the point $B_k = (k \cdot \Delta, y_k) = (k \cdot \Delta, -2f(k \cdot \Delta))$ is moved horizontally until it lies on the line from A_{k-1} to $(k \cdot \Delta, 0)$. In this position, we call the point $A_k = (x_k, y_k)$.

Our concern is to determine the position of the points A_k. We already know the corresponding y-values $y_k = -2f(k\Delta)$. It is more difficult to determine the x-values. The horizontal line through A_k, which also passes through B_k, intersects the "predecessor triangles" on the way to the y-axis. The individual sections δ_i in the respective triangles can be easily determined with similarity arguments (see Fig. 5.17). We have

$$\frac{\delta_0}{y_0 - y_k} = \frac{\Delta}{y_0} \text{ i.e. } \delta_0 = \Delta - \frac{\Delta y_k}{y_0}.$$

Next, we then obtain

$$\frac{\delta_1}{y_1 - y_k} = \frac{\Delta}{y_1} \text{ or } \delta_1 = \Delta - \frac{\Delta y_k}{y_1}$$

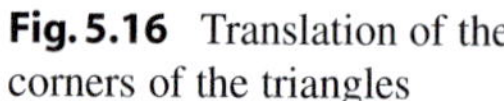

Fig. 5.16 Translation of the corners of the triangles

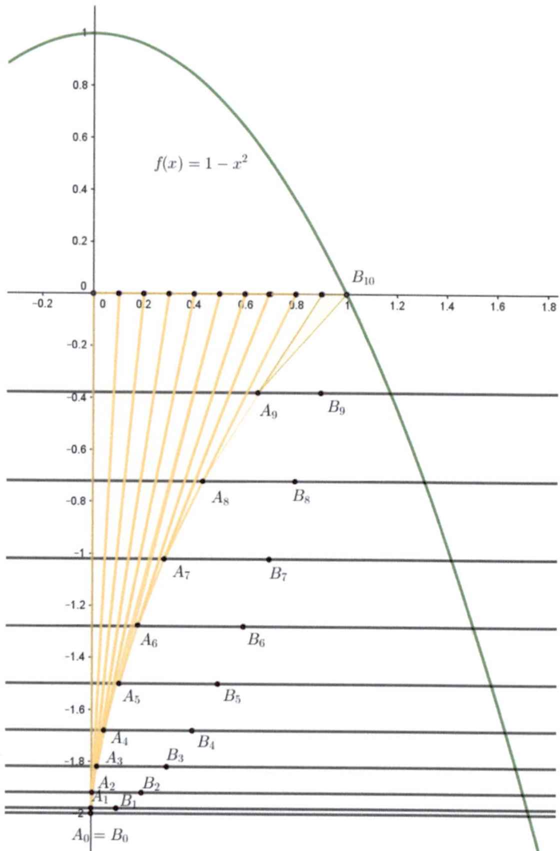

and in general

$$\frac{\delta_i}{y_i - y_k} = \frac{\Delta}{y_i} \quad \text{i.e.} \quad \delta_i = \Delta - \frac{\Delta y_k}{y_i},$$

(see Fig. 5.17). Here, $y_i - y_k$ represents the height in the smaller, darker triangle, while y_i represents the height in the larger triangle. δ_i and Δ then represent the corresponding parallel bases in the triangles. If we then add up all these contributions from δ_0 to δ_{k-1}, we obtain the desired x-value

$$x_k = k \cdot \Delta - \left(\frac{\Delta y_k}{y_0} + \frac{\Delta y_k}{y_1} + \cdots \frac{\Delta y_k}{y_{k-1}} \right) = \Delta \cdot k - y_k \sum_{i=0}^{k-1} \frac{\Delta}{y_i}.$$

Here we see that the expression $\frac{\Delta y_k}{y_0} + \frac{\Delta y_k}{y_1} + \cdots \frac{\Delta y_k}{y_{k-1}}$ exactly represents the length of the shift when B_k is moved to the position A_k.

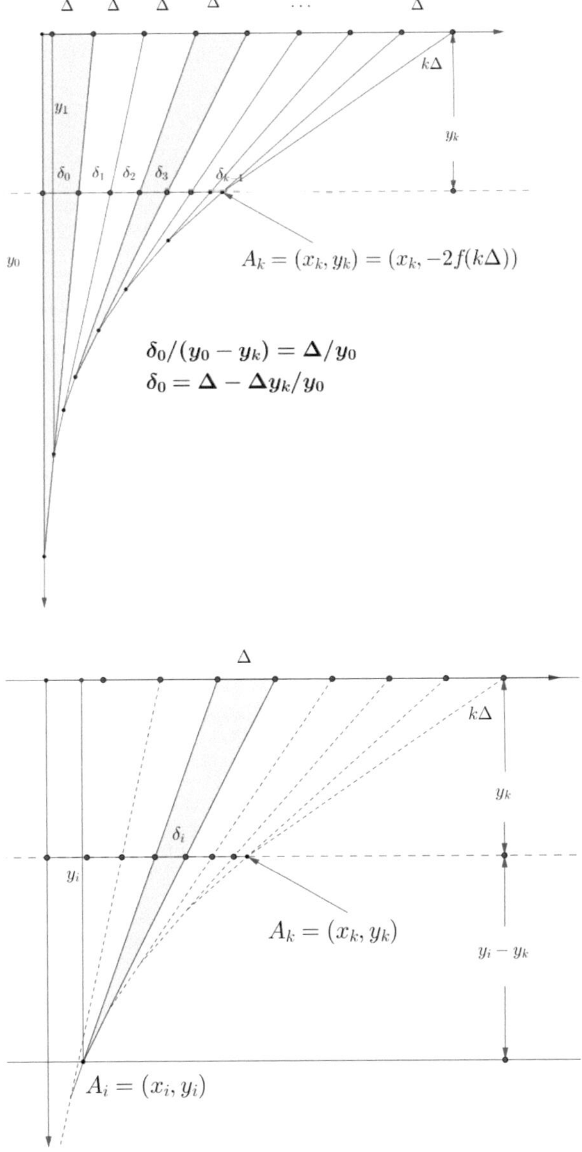

Fig. 5.17 Computation of the position of the points on the curve

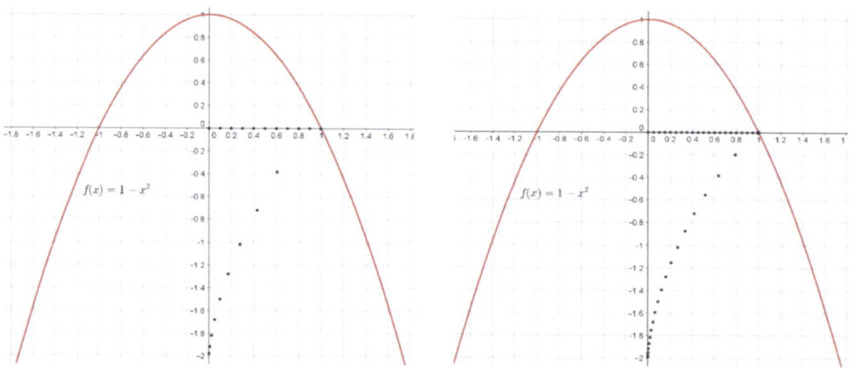

Fig. 5.18 The sequence of points for $n = 10$ and $n = 20$

We are now interested in the sequence of points A_i and conjecture that these points lie on an "underlying" curve and try to find this curve.

In Fig. 5.18, we see the sequence of points that belong to the same function $f(x) = 1 - x^2$ for the values $n = 10$ and $n = 20$. The pictures give us the impression that the sequences of points actually belong to an underlying curve. Now we increase the number n of subintervals but still consider the same point $s \in [0, b]$ in the interval. This means that we let n and k grow uniformly towards infinity

$$k\frac{b}{n} = k \cdot \Delta = s$$

and ask for the corresponding point $(x(s), y(s))$ on the curve. We obtain:

$$x(s) = \lim_{n\to\infty}\left(k \cdot \Delta - 2f(k \cdot \Delta)\sum_{i=0}^{k-1}\frac{\Delta}{2f(i \cdot \Delta)}\right) = s - f(s)\int_0^s \frac{dt}{f(t)}.$$

Here we have interpreted the expression $\lim_{k\to\infty} \sum_{i=0}^{k-1} \frac{\Delta}{f(i\cdot\Delta)}$ as the integral of the function $\frac{1}{f}$ over the interval $[0, s]$. In summary, we can say: Let $f(x) \neq 0$ be a continuous (monotonically decreasing) function. The limit curve that arises when, in the above-mentioned process, Riemann rectangles are exchanged for Riemann triangles, and their vertices are pushed together, we call *the associated fringe curve*. With this, we have found our first result on the Fringe Method.

Theorem 1 Let $f(x) \geq 0$ be a continuous, monotonically decreasing function above the x-axis over the interval $[0, a]$. The associated "fringe curve" below the x-axis is given by.

$$x(s) = s - f(s)\int_0^s \frac{dt}{f(t)} \text{ and } y(s) = -2f(s). \tag{5.1}$$

If $f(a) - 0$ then the area under the original curve equals the area over the fringe curve.

Remark 5.1
It is surprising that the integral of the reciprocal function $\frac{1}{f}$ appears here. Due to the complexity of the expression for $x(s)$ and the simplicity of the expression for $y(s)$, we will often consider inverse functions, when studying the fringe curve.

The following examples are intended to explain the method and provide insight into the details.

5.9 Examples

Example 1 $f(x) = (1-x)^2$.

If we use (5.1), we obtain.

$$x(s) = s - (1-s)^2\int_0^s \frac{dt}{(1-t)^2} = s - (1-s)^2\left[\frac{1}{1-t}\right]_0^s = s^2.$$

However, because $y(s) = -2(1-s)^2$, we get $y = -2\left(1-\sqrt{x}\right)^2 = 4\sqrt{x} - 2 - 2x$ (see Fig. 5.19).

We set $A = \int_0^1 (1-x)^2 dx$. Then also $A = \int_0^1 x^2 dx$ holds. Now $\sqrt{x}$ is the inverse function of x^2 and $\int_0^1 \sqrt{x}dx = 1 - A$ holds. With this, we get $A = \int_0^1 (1-x)^2 dx = -\int_0^1 \left(4\sqrt{x} - 2 - 2x\right)dx = 4(A-1) + 2\int_0^1 (1+x)dx = 4A - 1$, so $A = 1/3$ and we have determined the parabolic segment over the interval [0, 1].

However, we have used the integral formula $\int_0^s \frac{dt}{(1-t)^2} = \left[\frac{1}{1-t}\right]_0^s$ along the way. This relationship can be easily shown with the fan method from Chap. 6 (Sect. 6.7).

We will now consider an example where $f(a) \neq 0$ (see Fig. 5.20). In that case the area below the x-axis splits into two pieces. One is the area above the fringe curve and the other is a trapezoid. This is because the "last" triangle forms the right edge of the trapezoid when shifted to the left.

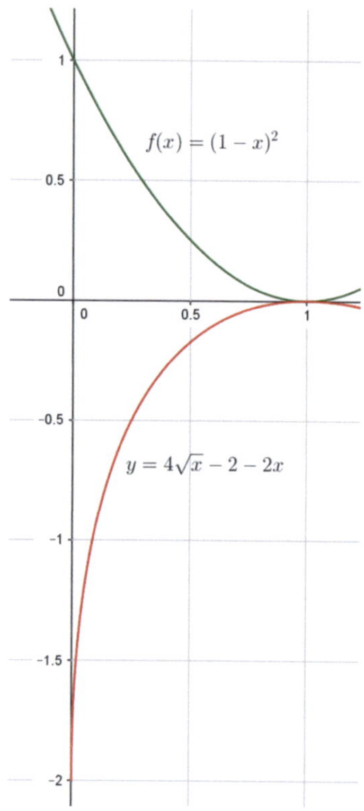

Fig. 5.19 The fringe curve of a parabola

Example 2 $f(x) = 1 - x$. The integral $\int \frac{dt}{f(t)}$ is the logarithm function. So, we can expect something new here.

$$x(s) = s - (1 - s)\int_0^s \frac{dt}{1 - t} = 1 - (1 - s) - (1 - s)\int_0^s \frac{dt}{1 - t}$$
$$= (1 - s)(-1 + \ln(1 - s)) + 1$$

Now $y(s) = -2(1 - s)$ so $1 - s = \frac{-y}{2}$ holds, and we have

$$x = -\frac{y}{2}\left(\ln\left(\frac{-y}{2}\right) - 1\right) + 1.$$

A linear function produces a logarithm with the Fringe Method, which is surprising at first. This also gives us the opportunity to gain new insights. Assuming we know the integral of the hyperbola $\int_0^s \frac{dt}{1-t} = -\ln(1 - s)$, which is important to find $x(s)$, then the Fringe Method can help us find $\int x \ln(x)dx$. As we saw above a trapezoid appears under the x-axis in Fig. 5.20. The short horizontal edge of the trapezoid then starts at the point $(x(a), y(a))$, where a is the right end of the integration interval.

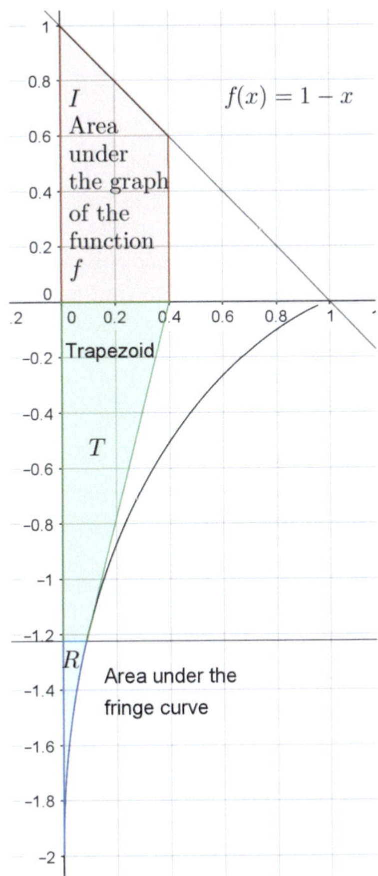

Fig. 5.20 The fringe curve of a linear function

For the original function f we get

$$I = \int_0^a f(x)dx = \int_0^a (1 - x)dx = a - \frac{a^2}{2}.$$

The area of the trapezoid T in Fig. 5.20 is

$$\begin{aligned} T &= (a + x(a))f(a) = (a + (1 - a)(-1 + \ln(a - 1)) + 1)(1 - a) \\ &= (1 - a)^2 \ln(1 - a) + 2a(1 - a). \end{aligned}$$

To compute the area R we will consider the reflected inverse function across the line $y = x$ and we get

$$R = \int_{-2}^{-2(1-a)} \left(\frac{-x}{2} \left(\ln\left(\frac{-x}{2} \right) - 1 \right) + 1 \right) dx.$$

Because $\int x \ln(x) dx$ only appears in the expression for R, the relationship $I = R + T$ can be used to find an expression for $\int x \ln(x) dx$. Thus, by applying the Fringe Method here, we have solved a new problem, namely finding $\int x \ln(x) dx$.

Example 3 The bell curve $f(x) = \frac{1}{1+x^2}$.

This curve (see Fig. 5.21) had already appeared in Leibniz's circle quadrature. The corresponding fringe curve can be described with (5.1) as

$$x(s) = s - \frac{s + \frac{s^3}{3}}{1 + s^2} = \frac{2s^3}{3(1 + s^2)} \text{ and } y(s) = \frac{-2}{1 + s^2}, \text{ so } s^2 = -\frac{y + 2}{y}.$$

Therefore

$$x = \frac{2s^3 y}{3(-2)} = -\frac{y}{3} \left(-\frac{(y + 2)}{y} \right)^{\frac{3}{2}} = \frac{1}{3} \sqrt{-\frac{y^2 (y + 2)^3}{y^3}} = \frac{1}{3} \sqrt{-\frac{(y + 2)^3}{y}}.$$

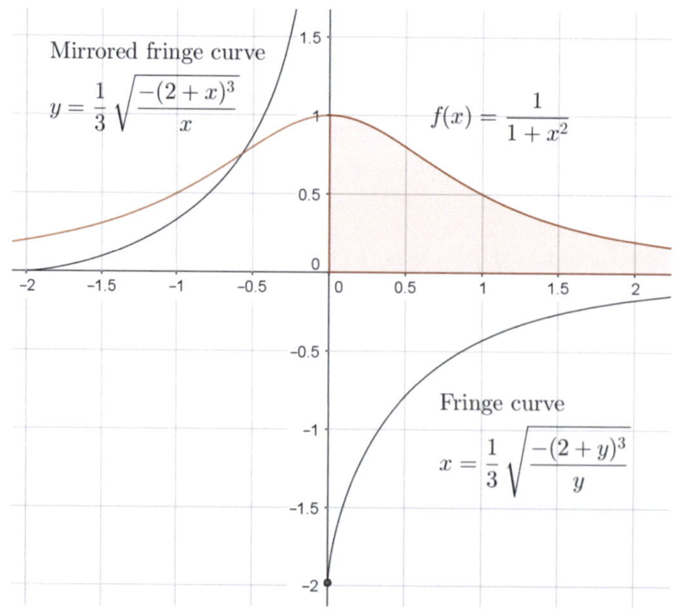

Fig. 5.21 The fringe curve of the Bell curve

Reflection about the mirror line $y = x$ gives us the curve $y = \frac{1}{3}\sqrt{-\frac{(2+x)^3}{x}}$ (see Fig. 5.21). Because it is well known that $\int_0^\infty \frac{dx}{1+x^2} = \frac{\pi}{2}$, which we can see in Chap. 6 about the fan method by Aage Bondesen, we get the non-trivial integral equation

$$\frac{\pi}{2} = \int_0^\infty \frac{dx}{1+x^2} = \frac{1}{3}\int_{-2}^0 \sqrt{\frac{-(2+x)^3}{x}}dx.$$

Here we consider an improper integral and the function graph approaches the x-axis for increasing x-values. Therefor we can again disregard the trapezoid.

Example 4 Let's choose a quarter circle $f(x) = \sqrt{1-x^2}$, so we obtain with (5.1).

$$x(s) = s - \sqrt{1-s^2}\int_0^s \frac{1}{\sqrt{1-t^2}}dt = s - \sqrt{1-s^2}\arcsin(s)$$

Now we have $y(s) = -2\sqrt{1-s^2}$ or $s = \sqrt{1-\left(\frac{y}{2}\right)^2}$. Therefore, we get

$$x = \sqrt{1-\left(\frac{y}{2}\right)^2} + \frac{y}{2}\arcsin\left(\sqrt{1-\left(\frac{y}{2}\right)^2}\right)$$

If we reflect the function graph about the line $y = x$, we get

$$y = \sqrt{1-\left(\frac{x}{2}\right)^2} + \frac{x}{2}\arcsin\left(\sqrt{1-\left(\frac{x}{2}\right)^2}\right).$$

Because the quarter circle has an area of $\frac{\pi}{4}$, we know that

$$\int_{-2}^0 \left(\sqrt{1-\left(\frac{x}{2}\right)^2} + \frac{x}{2}\arcsin\left(\sqrt{1-\left(\frac{x}{2}\right)^2}\right)\right)dx = \frac{\pi}{4}.$$

Here we can interpret the function $y = \sqrt{1-\left(\frac{x}{2}\right)^2}$ as an ellipse with semi-axes 2 and 1 (see Fig. 5.22). The area of the entire ellipse is 2π. In our case,

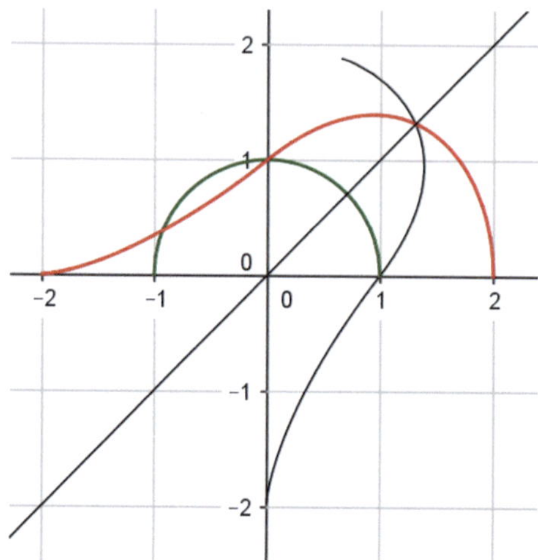

Fig. 5.22 The fringe curve of a quarter circle

only the quarter ellipse is relevant, whose area is then $\frac{\pi}{2}$. Therefore, we have $\int_{-2}^{0} \sqrt{1-\left(\frac{x}{2}\right)^2}dx = \frac{\pi}{2}$ and we get

$$\int_{-2}^{0} \frac{x}{2}\arcsin\left(\sqrt{1-\left(\frac{x}{2}\right)^2}\right)dx = -\frac{\pi}{4}.$$

The graph of the "function" $h(x) = \pm\frac{x}{2}\arcsin\left(\sqrt{1-\left(\frac{x}{2}\right)^2}\right)$, which only forms a part of the fringe curve, is interesting in itself. It forms a lemniscate-like figure (see Fig. 5.23) with a total area of π.

5.10 Properties of the Fringe Curve

Theorem 2 Given a continuous, monotonically decreasing function $f(x) > 0$ for $0 \le x < a$ and $f(a) = 0$ over the interval $[0, a]$ and the curve $g(x) = -2f(x)$, on which the reflected triangle vertices lie. The corresponding fringe curve divides the area under the function graph of g over the interval $[0, a]$ into two parts of equal area.

Proof Since the area over g (red) is twice as large as that under f (green) and the area under the fringe curve is the same as that under f, the remaining area must also be equal (see Fig. 5.24).

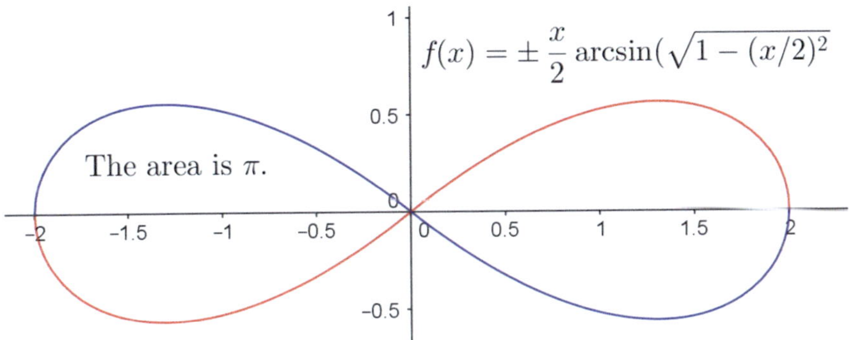

Fig. 5.23 Lemniscate with area π

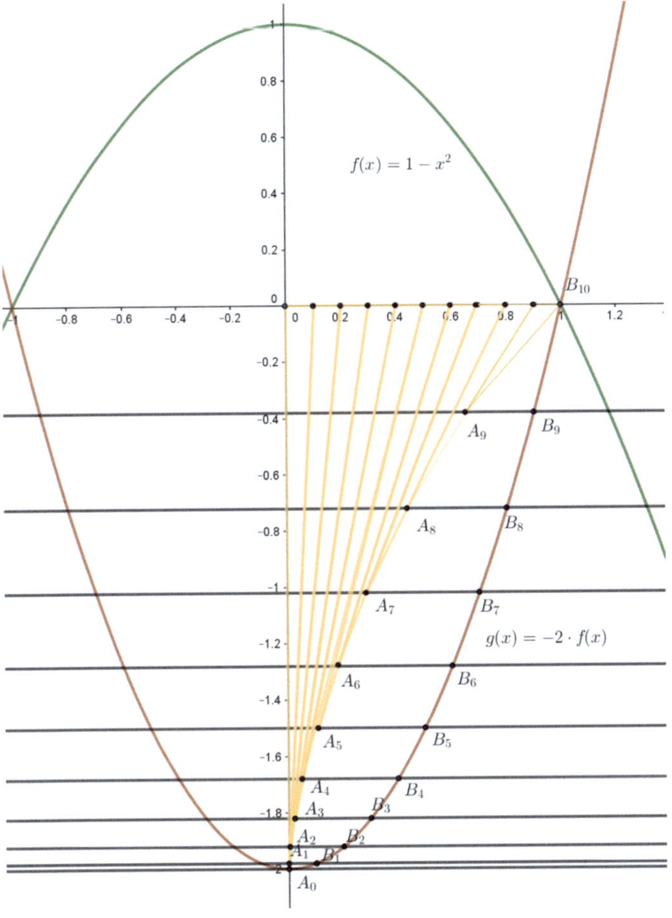

Fig. 5.24 Division of the area

Theorem 3 Given a continuous, monotonically decreasing function $f(x) > 0$ over the interval $[0, a]$ and the corresponding fringe curve

$$x(s) = s - f(s)\int_0^s \frac{dt}{f(t)} \text{ and } y(s) = -2f(s) \text{ for } s \in [0, a].$$

The tangent through a point $P = (x(s), y(s))$ on the fringe curve intersects the x-axis at $(s, 0)$.

Proof

First, we calculate the slope of the tangent at point P using (5.1).

$$\frac{dy}{dx} = \frac{\frac{dy}{ds}}{\frac{dx}{ds}} = \frac{-2f'(s)}{1 - f'(s)\int_0^s \frac{dt}{f(t)} - f(s)\frac{1}{f(s)}} = \frac{-2f'(s)}{-f'(s)\int_0^s \frac{dt}{f(t)}} = \frac{2}{\int_0^s \frac{dt}{f(t)}} = m$$

The equation of the tangent is then $y - y(s) = m(x - x(s))$. We are interested in the intersection of the tangent with the x-axis, so $y = 0$. This gives $-\frac{y(s)}{m} = x - x(s)$ or

$$\frac{-(-2f(s))}{2}\int_0^s \frac{dt}{f(t)} = x - \left(s - f(s)\int_0^s \frac{dt}{f(t)}\right),$$

thus $x = s$.

Remark 5.2
To understand how this result comes about, we can consider the three points A_{k-1}, A_k and $(k\Delta, 0)$ in the initial presentation of the Fringe Method, see Fig. 5.16. These three points lie on a straight line because the triangle with vertex B_k was pushed onto the preceding triangle. In the limit process, this line becomes the tangent because A_{k-1} and A_k are neighboring points on the curve and are infinitely close to each other. The intersection with the x-axis $(k\Delta, 0)$ is the point $(s, 0)$, because $k\Delta = s$. This shows the result without having to use calculus, as above.

Now we come to our main result.

Theorem 4 The Resection Method is the inverse of the Fringe Method.

Proof We start with the function $f(x)$ (continuous and monotonically decreasing) above the x-axis. The points of the curve can be described by $P = (s, f(s))$. Assuming we have already found the corresponding fringe curve, a point on the fringe curve then has the coordinates $Q = (x(s), -2f(s))$. We now apply the Resection Method to this curve, which means that we have to construct the tangent at the point Q, and intersect it with the x-axis. But this is exactly what we did in Theorem 3. The

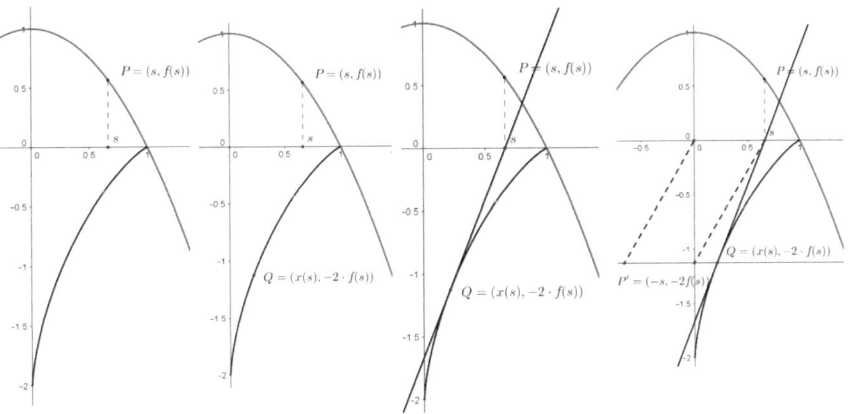

Fig. 5.25 Steps in the proof that the resection method is the inverse of the fringe method

intersection is $(s, 0)$. The length, s, is now used to find the point P ' on the new curve, the resect, at the same height as Q, (see Fig. 5.25). This is then $P' = (-s, -2f(s))$. Thus, the resection curve is essentially identical to the initial curve $(s, f(s))$. We just need to reflect it around the origin and reduce it by a factor of 2. This makes sense, as the Resection Method gives us an area that is twice as large as the original one.

5.11 Leibniz's Resection Method in the Light of the Fringe Method

The Fringe Method has a consistently geometric interpretation. Riemann rectangles are exchanged for triangles, which are sheared until a new connected area is formed. Now, the Fringe Method and the Resection Method are inverses, so we can hope that the Resection Method itself allows a similar geometric interpretation, so that we can better understand the Resection Method.

We would like to undertake this endeavor, namely to give the Resection Method a geometric interpretation.

We want to determine the area between the arc $OQPA$ and the line AO using the Resection Method. A typical triangle PQO is part of this area. We now shift the sharp corner O on a parallel to PQ, until the side through P becomes horizontal. We obtain the triangle PQJ, which has the same area as PQO. We now translate the triangle PQJ horizontally until it lies next to the y-axis and fill it up to a rectangle with double the area (see Fig. 5.26). If we do this for all subtriangles that fill the segment, we fill the area under the resection curve to the left of the y-axis. The neighboring points P and Q come infinitely close together in the limit case and their connecting line then becomes the tangent, so that the baseline PJ in the triangle, which later comes to lie to the left of the y-axis, is equal

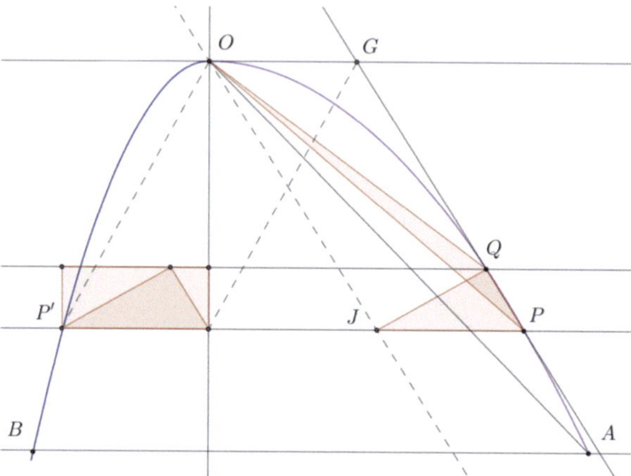

Fig. 5.26 Resection Method

to the tangent segment OG. Now the triangles of the form PQO fill the segment between the curve and the line AO, while the small rectangles fill the corresponding area between the y-axis, the new resection curve and the horizontal closing lines through A and B. Thus, the area under the new resection curve is twice as large as that under the old one, and we have found a geometric interpretation of the Resection Method à la the Fringe Method.

Here we cite Leibniz (2016, S. 35) himself with some small adaptations:

> "Under these assumptions, it is clear from Theorem 1 that the double of one triangle is the corresponding rectangle and that the double of the next triangle is the next rectangle, and so on for any remaining areas. Thus, the sum of any number of rectangles of this kind, or the step-shaped area, will be twice as large as the sums of all such triangles, or as the inscribed polygon."

5.12 Summary

In the geometric interpretation of Leibniz's proof, we can see how the shift of the triangle vertices along lines parallel to the base lines ensures area equality. In this representation, the connection to the Fringe Method is also more apparent, as it can be seen as a shift of the aforementioned triangle vertices. Both methods, the Resection Method and the Fringe Method, are based on simple elementary geometric and particularly intuitive arguments, which can easily appeal to students, but can also establish a connection between tangents and areas, thus preparing for work with the fundamental theorem of calculus.

Overall, engaging with Leibniz's Resection Method can awaken a historical awareness of mathematics in students, as it describes a historical method that had great impact at the time of its development. In addition, students also see that other methods have surpassed the Resection Method and are now at the center of introductory lectures.

Then again, it is interesting to see that these old methods can still be developed today, for example, through the development of the Fringe Method as the inverse of the Resection Method.

Geometrically, this inverse seems easily understandable, as it involves displacement processes in one direction for the Resection Method and the other direction for the Fringe Method. However, algebraically, in terms of the resulting formulas (5.1), the connection is not so easy to see, and the appearance of the integral of the reciprocal function is particularly surprising.

5.13 Afterword: The Fringe Method for Parameter Curves

Sometimes our functions can be given in a parametric representation. Here, the coordinate is not given as a function of the ordinate, but both coordinates, the x-coordinate and the y-coordinate, are given as functions of a third variable. If the original monotonically decreasing function is represented by $x = x(q)$ and $y = y(q)$, then the fringe curve takes the form:

$$\overline{x} = x(q) - y(q) \int_{x^{-1}(0)}^{q} \frac{x'(t)}{y(t)} dt \text{ and } \overline{y} = -2y(q). \tag{5.2}$$

Example For the quarter circle $x = \cos(q), y = \sin(q)$ for $0 \leq q \leq \frac{\pi}{2}$ we get.

$$\begin{aligned}
\overline{x} &= x(q) - y(q) \int_{x^{-1}(0)}^{q} \frac{x'(t)}{y(t)} dt \\
&= \cos(q) - \sin(q) \int_{\arccos(0)}^{q} \frac{-\sin(t)}{\sin(t)} dt \\
&= \cos(q) + \sin(q) \int_{\frac{\pi}{2}}^{q} dt = \cos(q) + \sin(q)\left(q - \frac{\pi}{2}\right), \overline{y}(q) = -2\sin(q).
\end{aligned}$$

The integral $\int \frac{x'(t)}{y'(t)} dt = \int (-1) dt$ is particularly simple. This also shows that it is sometimes worthwhile to use the parameter representations of the curves (see Fig. 5.27).

If we write $\cos(q) = \sqrt{1 - \sin^2(q)} = \sqrt{1 - \left(\frac{\overline{y}}{2}\right)^2}$ and $q - \frac{\pi}{2} = \arcsin\left(\frac{-\overline{y}}{2}\right) - \frac{\pi}{2}$, we get the same result as in Example 4, where we already had treated the circle in

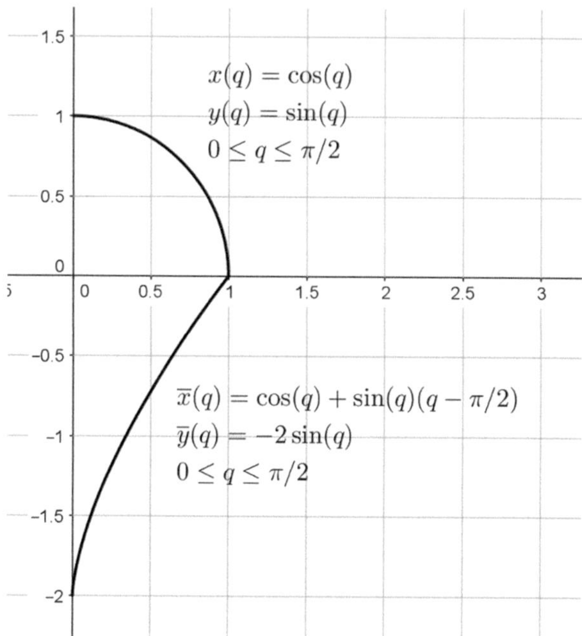

Fig. 5.27 Fringe curve belonging to a quarter of a circle (parameter representation)

the representation $y = \sqrt{1 - x^2}$. The result is

$$x = \sqrt{1 - \left(\frac{y}{2}\right)^2} + \frac{y}{2}\arcsin\left(\sqrt{1 - \left(\frac{y}{2}\right)^2}\right).$$

In the case of parameter curves we can still show that the Resection Method and the Fringe Method are inverse methods of each other. The differentiable (monotonically decreasing) starting curve above the x-axis is given by $x = x(q)$ and $y = y(q)$. If we again can show that the tangent at the curve point $(\overline{x}(q), \overline{y}(q))$ intersects the x-axis at $(x(q), 0)$, then the result follows similarly as above.

With (5.2) we get the following for the slope of the tangent at the curve point $(\overline{x}(q), \overline{y}(q))$

$$\frac{d\overline{y}}{d\overline{x}} = \frac{\frac{d\overline{y}}{dq}}{\frac{d\overline{x}}{dq}} = \frac{-2y'(q)}{x'(q) - y'(q)\int_{x^{-1}(0)}^{q} \frac{x'(t)dt}{y(t)} - y(q) \cdot \frac{x'(q)}{y(q)}} = \frac{-2y'(q)}{-y'(q)\int_{x^{-1}(0)}^{q} \frac{x'(t)dt}{y(t)}}$$

$$= \frac{2}{\int_{x^{-1}(0)}^{q} \frac{x'(t)dt}{y(t)}} = m$$

The equation of the tangent then has the form $y - \overline{y}(q) = m(x - \overline{x}(q))$. We are now interested in the intersection point of the tangent with the x-axis, i.e., $y = 0$. Then we have

$$\frac{-(-2y(q))}{2} \int_{x^{-1}(0)}^{q} \frac{x'(t)dt}{y(t)} = x - \left(x(q) - y(q) \int_{x^{-1}(0)}^{q} \frac{x'(t)dt}{y(t)} \right),$$

so, $x = x(q)$. This again means that the Fringe Method is the inverse of the Resection Method, even if the starting curve is given as a parameter curve.

Bibliography

Kirfel, C. (2023): Die Fransenmethode zur Berechnung von Flächen, In: IDMI-Primar Goethe-Universität Frankfurt (Hrsg.) WTM, 56. Jahrestagung der Gesellschaft für Didaktik der Mathematik, Beiträge zum Mathematikunterricht 2022, pp. 917–920. https://doi.org/10.37626/GA9783959872089.0

Kirfel, C. (2024): Die geometrische Umkehrung der Resektenmethode von Leibniz, in: Vom Mittelater in die Moderne – Episoden aus der Mathematikgeschichte, Tagungsband der Konferenz für die Geschichte der Mathematik 2023, Leipzig, WTM-Verlag, Münster, pp. 103–116.

Leibniz, G. W. (2016): De quadratura arithmetica circuli ellipseos et hyperbolae. Herausgegeben v. Eberhard Knobloch. Klassische Texte der Wissenschaft. Berlin, Heidelberg: Springer Spektrum 2016.

Probst S. (2016): "Leibniz und Roberval", in: Wenchao Li et al. (Hrsg.), "Für unser Glück oder das Glück anderer". Vorträge des X. Internationalen Leibniz-Kongresses Hannover, 18.-23. Juli 2016, Hildesheim, 2016, Vol IV, pp. 183–189.

Probst, S. (2017): Leibniz as Reader and Second Inventor: The Cases of Barrow and Mengoli, in G.W. Leibniz, Interrelations Between Mathematics and Philosophy, Springer, 2017, pp. 111–134.

Ullrich, P. (2017): Das Manuskript von Leibniz aus dem Jahre 1676 über Infinitesimalrechnung, MU der Mathematikunterricht, Jahrgang 63, Vol 4, August 2017, pp. 18–28.

The Fan Method

6

6.1 Introduction

In this chapter, we present an integration method that originates from Fermat, but has also found use in a modern presentation in Roger B. Nelsen's book, Proofs Without Words II (Nelsen 2000). In the modern presentation, which is heavily inspired by geometry, we will see how Riemann rectangles, which approximate the area under a curve, are transformed into triangles. In this way, a new figure is created with the same area, and we can often draw interesting conclusions (see also Aslaksen and Kirfel 2024; Kirfel 2020).

Only quite simple arguments, similarity and, as already mentioned, Riemann rectangles are used. The theory of the fan method is built up through a series of examples. In the end, we will see how the method can also be extended to three-dimensional figures, the volume of which we can then determine.

In the chapter on Fermat's integration methods, we have already learned two methods that Pierre de Fermat (1896) developed around 1650 to calculate areas under curves. In addition, Fermat's writing contains another part, which is much more difficult to access. This part deals with much more complicated integration problems. Among other things, Fermat succeeds in integrating the so-called Witch of Agnesi (see also Chap. 5). This is the function

$$y = \frac{1}{1 + x^2},$$

Which is called a bell curve (see Fig. 6.1). The name "Witch of Agnesi" is based on a translation error. Maria Gaetana Agnesi, an Italian mathematician, studied this curve extensively in 1748. As mentioned above, the curve also appears in Fermat's work. Agnesi named it "averisera" (inverse curve). This was mistakenly translated in her original work to "avversiera", which means "witch" or "wife of the devil", which then became the "Witch of Agnesi" in an English translation [Agnesi].

© The Author(s), under exclusive license to Springer-Verlag GmbH, DE, part of Springer Nature 2026

C. Kirfel, *Side Paths in the History of Mathematics*, Mathematics Study Resources 21,
https://doi.org/10.1007/978-3-662-72918-2_6

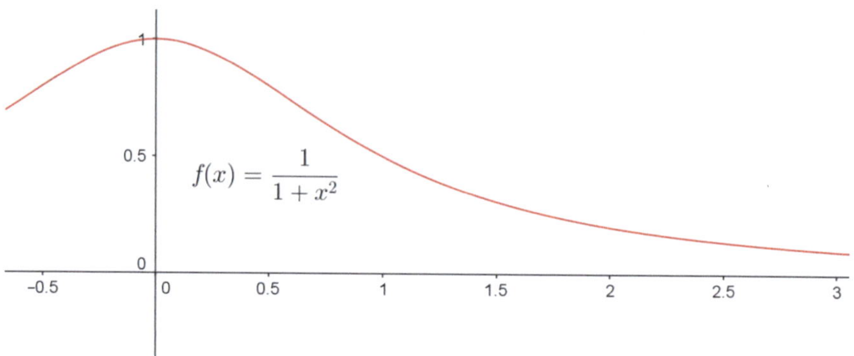

Fig. 6.1 "The Witch of Agnesi"

Fermat's approach is algebraic in nature. He uses two variable substitutions and can thus determine the area under the bell curve. His arguments are quite difficult to access and are not suitable for the present presentation, see also Remark 6.10. We have therefore decided to resort to a geometric integration of the mentioned curve, which can be found in Roger B. Nelsen's "Proofs Without Words II" (2000). Roger B. Nelsen collects a number of mathematical results with visual proofs that do not require further explanation with many words. There is also a beautiful illustration of the following formula (2000, p. 63)

$$\int_0^a \frac{1}{(1+x^2)} dx = \arctan(a),$$

which goes back to Aage Bondesen (see Fig. 6.2).

6.2 The Integration of the Witch of Agnesi

The Riemann integral under the graph of the function $f(x) = \frac{1}{2(1+x^2)}$, can be understood as the limit of the sum of the Riemann rectangles. Such a Riemann rectangle *CDBF* can be seen in Fig. 6.2. The width of the rectangle or the distance of the corners *C* and *D* on the *x*-axis is Δ. We now connect the corners *C* and *D* with the point $R = (0, -1)$ and obtain the triangle *RCD* with the area $\Delta \cdot \frac{1}{2}$, because the height is just one unit. Now comes the trick: We reduce the triangle *CDR* to the size *KJR*, where *K* lies on a unit circle, which is shifted down by one unit in order to create a new area that does not overlap with the old one (Fig. 6.2). The reduction factor when transitioning from *CDR* to *KJR* is

$$\frac{KR}{CR} = \frac{1}{\sqrt{1+x^2}}.$$

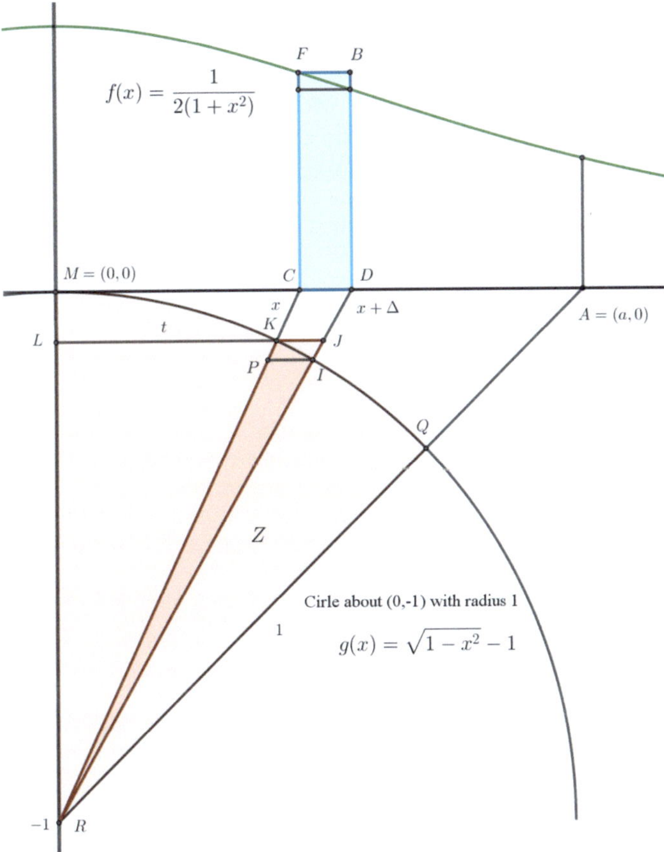

Fig. 6.2 Integration of "The Which of Agnesi"

Here we have used the Pythagorean theorem in the triangle *RCM*. To calculate the area of the reduced triangle *KJR*, we have to multiply the original area $\Delta \cdot \frac{1}{2}$ of the triangle *CDR* by the square of the reduction factor KR/CR, because in area calculations the "scale" appears twice, once for the baseline and once for the height of the triangle. Thus, the triangle *KJR* has the area

$$\frac{\Delta}{2(1+x^2)}.$$

This corresponds exactly to the area of the Riemann rectangle *CDBF* under the graph of the function *f*. Since we started with the function $f(x) = \frac{1}{2(1+x^2)}$ and not $f(x) = \frac{1}{(1+x^2)}$ the new area lies completely under the *x*-axis. This is not logically necessary and later on we will not make this adjustment, but for now it makes the figures easier to understand. We now add up the areas of the Riemann rectangles under the function graph over the interval $[0, a]$. This gives us $\int_0^a \frac{1}{2(1+x^2)}dx$. On the other hand, if we sum the "wedge-shaped circle pieces", we get the circle sector Z, which corresponds to the angle $\sphericalangle QRM$. The tangent of this angle is a, which we can read off from the x-axis. We denote both the circle sector and its area with Z. Since the full circle has the circumference 2π and the area π, the area of the sector Z corresponds to half the angle, so

$$\frac{\sphericalangle QRM}{2\pi} = \frac{Z}{\pi}$$

and thus we get $\int_0^a \frac{dx}{2(1+x^2)} = Z = \frac{\sphericalangle QRM}{2} = \frac{\arctan(a)}{2}$, so

$$\int\limits_0^a \frac{dx}{1+x^2} = \arctan(a).$$

This completes the proof. In particular, we also have $\int\limits_0^\infty \frac{dx}{1+x^2} = \frac{\pi}{2}$.

We can imagine the process like this. The Riemann rectangles above the x-axis are transformed into area-equivalent "Riemann triangles". These then form a fan-like figure. In this case, the limit of the fan-like figure is a circle sector, whose area was easy to calculate. This leads us to the integral of the bell curve.

We might now wonder how we knew we had to use a unit circle below the x-axis. So, we are interested in understanding where the equation of the fan curve comes from. To do this, we examine the coordinates of the point $K = (t, g(t))$. We will call this curve g the fan curve. We compare the vertical catheti, the horizontal catheti, and the hypotenuses of the two similar triangles RMC and RLK. This gives us

$$\frac{g(t)+1}{1} = \frac{t}{x} = \frac{1}{\sqrt{1+x^2}}.$$

Here we have $t = \frac{x}{\sqrt{1+x^2}}$ and thus $t^2\left(1+x^2\right) = x^2$ or $x^2\left(1-t^2\right) = t^2$, i.e. $x^2 = \frac{t^2}{1-t^2}$ and thus

$$g(t)+1 = \frac{1}{\sqrt{1+\frac{t^2}{1-t^2}}} = \sqrt{1-t^2},$$

which just represents a shifted unit circle.

We now leave the specific example of the "Witch of Agnesi" and generalize the idea presented. It is about establishing a connection between two integrals. One corresponds to the area under a function graph above the x-axis, while the other represents the fan-like area below the x-axis. The Riemann rectangles above the x-axis are identified with triangles of the same area, with vertex at the point

$R = (0, -1)$. This creates a fan-like area. In our first example, this fan-like area was easy to calculate. In later examples, the fan-like areas can be represented by other integrals. But each time, by converting Riemann rectangles into Riemann triangles we uncover interesting connections between integrals of different curves. We call the method the "fan method".

6.3 The Linear Function

Surprisingly, the fan curve of a linear function is not itself a linear function or any simple mathematical function, but unexpectedly a cube root. We now want to investigate this further and choose our linear function $y = \frac{x}{2}$ above the x-axis and ask for the corresponding fan function $g(t)$ below the x-axis. As above, we have also here identified Riemann rectangles with Riemann triangles. Also here, the triangles CMR and PQR are similar (see Fig. 6.3). The ratio of the horizontal to the vertical leg in both triangles is therefore the same and we can write

$$\frac{x}{1} = \frac{t}{QR} = \frac{t}{g(t) + 1}. \tag{6.1}$$

In addition, we know that the Riemann rectangles and the corresponding Riemann triangles have the same area. The reduction factor used is again t/x. With this, we can determine the fan curve $g(t)$ by equating the mentioned areas. This gives us

$$\Delta \cdot \frac{x}{2} = \Delta \cdot f(x) = \frac{\Delta}{2} \cdot \left(\frac{t}{x}\right)^2,$$

which in our case means $x^3 = t^2$ or $x = \sqrt[3]{t^2}$. Because of (6.1), this gives us again

$$g(t) + 1 = \frac{t}{x} = \frac{t}{\sqrt[3]{t^2}} = \sqrt[3]{t}.$$

Here we have again exploited the equality of the ratios of the vertical and horizontal legs. For the endpoint $A = (a, 0)$ of the integration interval, we always have,

$$\frac{a}{1} = \frac{T}{g(T) + 1},$$

because the corresponding triangles AMR and SLR are also similar.

Now we again add all Riemann rectangles over the interval $[0, A]$. The area of the triangle under the function f above the x-axis, which is $a^2/4$, is then equal to the curvilinearly bounded area Z, which lies between the graph of the function $y = \sqrt[3]{x} - 1$ and the line through R and $A = (a, 0)$.

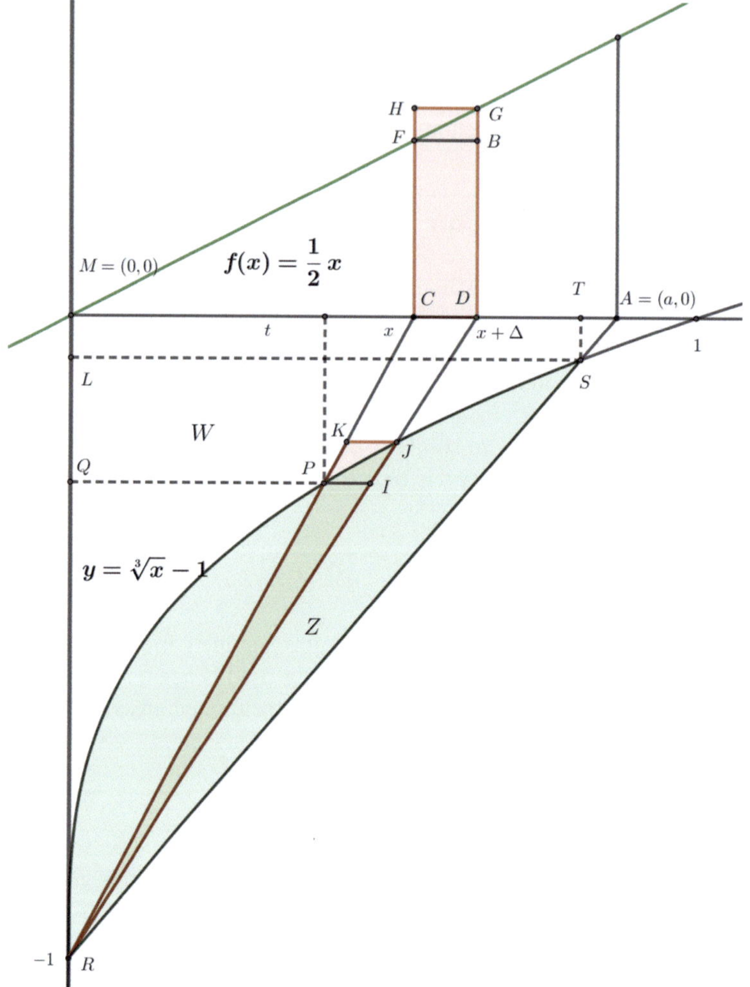

Fig. 6.3 The linear function

The white area to the left of the fan curve can be considered as an integral of the inverse function of the fan curve, i.e., x^3 (shifted downwards). For this white area W between R and L, we have

$$W = \int_0^{\sqrt[3]{T}} x^3 dx,$$

because the distance between R and L is exactly $\sqrt[3]{T}$. For $0 < a < 1$ this results in

$$\frac{a^2}{4} = \int_0^a \frac{x}{2} dx = Z = \text{Area}(RSL) - \int_0^{\sqrt[3]{T}} x^3 dx = \frac{T\sqrt[3]{T}}{2} - \int_0^{\sqrt[3]{T}} x^3 dx,$$

where $\frac{T\sqrt[3]{T}}{2}$ represents the area of the triangle RLS. Because of (6.1), we have

$$a = \frac{T}{\sqrt[3]{T}} = T^{\frac{2}{3}} \quad \text{i.e.} \quad \sqrt{a} = \sqrt[3]{T}$$

and the area of the triangle RSL is $\frac{T\sqrt[3]{T}}{2} = \frac{a^2}{2}$. In total, we get

$$\frac{a^2}{4} = \frac{a^2}{2} - \int_0^{\sqrt{a}} x^3 dx, \text{ i.e. } \int_0^{\sqrt{a}} x^3 dx = \frac{a^2}{4}, \text{ or } \int_0^{V} x^3 dx = \frac{V^4}{4}.$$

This confirms the known integration formula for power functions of the third degree from analysis. We will also use this technique in the upcoming examples.

Remark 6.1
This argument was carried out for $a < 1$. For $a > 1$, the fan curve will cross the x axis. However, this is not problematic if we replace the function g with a function $g(x) = \sqrt[3]{x} - c$ with $c > a$. In this way, we can be sure that $g(x)$ lies below the x-axis and that the figures do not overlap. However, this is not essential for the method itself. It simply makes the visualization easier. In this way we see that the integral formula is valid for all values of A.

With the two examples, the Witch of Agnesi and the linear function, fresh in mind, we now want to try to generalize our observations and give the fan method a fixed form. If now $f(x)$ is a continuous function above the x-axis, and $g(t)$ the corresponding fan curve below the x-axis, where the Riemann rectangles have been replaced by Riemann triangles, then we have the following two equations

$$\frac{x}{1} = \frac{t}{g(t) + 1}, \tag{6.1}$$

which represents the similarity of the triangles RMC and RQP, and

$$\Delta \cdot f(x) = \frac{\Delta}{2} \cdot \left(\frac{t}{x}\right)^2, \tag{6.2}$$

which expresses the area equality of the Riemann rectangles and the triangles. In total, this gives us:

Theorem 1 Given a continuous function $f(x)$. The corresponding fan curve can then be described by

$$f\left(\frac{t}{g(t)+1}\right) = \frac{(g(t)+1)^2}{2}. \tag{6.3}$$

Remark 6.2
With this formula, we can find the fan curve g below the x-axis once f is given. The problem of finding the fan curve is thus solved once and for all. We will frequently apply this formula in the following examples.

Remark 6.3
However, because g is given by the implicit formula (6.3), we cannot always solve the equation for g and we then get a curve instead of a function. This will be the case in a series of the following examples (Sections 6.7, 6.8, 6.9 and 6.10).

6.4 The Parabola

After studying linear functions above, it is natural to also look at the quadratic case. Here too something astonishing happens. It turns out that the fan curve is also a quadratic function. But because both functions have different coefficients, namely ½ and 1, we manage to exploit the area equality, so that we can get a formula for the area under the parabola. This is an unusual stroke of luck.

We now consider a quadratic function $f(x) = \frac{x^2}{2}$ above the x-axis and ask for the corresponding fan curve $g(t)$ below the x-axis, where the Riemann rectangles have been replaced by Riemann triangles as before. Equation (6.3), $f\left(\frac{t}{g(t)+1}\right) = \frac{(g(t)+1)^2}{2}$, then gives us the following:

$$\frac{1}{2} \cdot \frac{t^2}{(g(t)+1)^2} = \frac{(g(t)+1)^2}{2} \quad \text{or} \quad g(t)+1 = \sqrt{t}.$$

As already anticipated above, this means that the fan curve itself is a root function, which is the inverse of a quadratic function.

If we now sum the Riemann rectangles under the parabola f over the interval $[0, a]$, the total area corresponds to the area Z between the graph of the fan curve $g(t) = \sqrt{t} - 1$ and the line RS. Now we apply the same "inversion trick" as above, by considering the white residual area to the left of the fan curve. Again, the area Z can be understood as the difference of the triangle area RSL and the white residual area W (see Fig. 6.4). This residual area can be seen as the area under a parabola

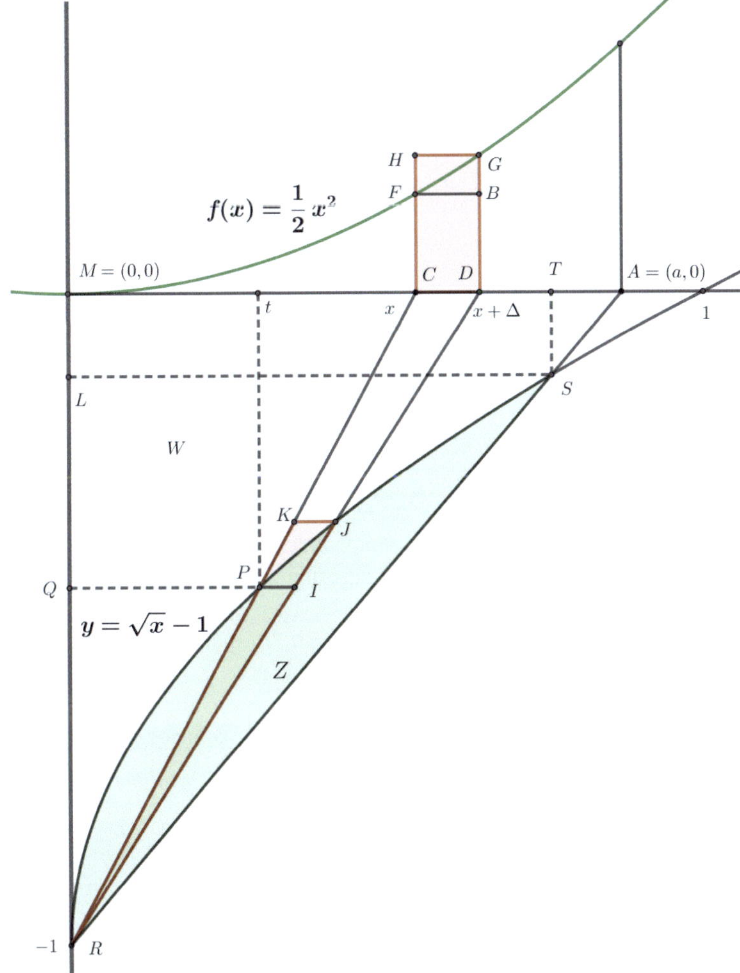

Fig. 6.4 The integration of the parabola

because the function g can be seen as the inverse function of a quadratic function and we have

$$W = \int_0^{\sqrt{T}} x^2 dx$$

Because of (6.1) we have $a = \frac{T}{g(T)+1} = \frac{T}{\sqrt{T}} = \sqrt{T}$. Thus, the area of the triangle *RSL* is equal to $\frac{T\sqrt{T}}{2} = \frac{a^3}{2}$. This gives us

$$\int_0^a \frac{x^2}{2} dx = Z = \frac{a^3}{2} - \int_0^a x^2 dx.$$

In total, we then have

$$\frac{3}{2}\int_0^a \frac{x^2}{2} dx = \frac{a^3}{2}, \text{ i.e. } \int_0^a x^2 dx = \frac{a^3}{3}$$

and again we have succeeded in confirming a known result from analysis. We have managed to integrate the parabola.

In the preceding examples, our argument switched between the desired areas and the corresponding residual areas, which belong to the inverse functions. We would now like to refine and generally describe this technique, so that we can apply it in more complex situations.

We now consider the rectangle *MAPW*, which "frames" the graph of our function $f(x)$ over the interval $[0, a]$ (see Fig. 6.5). The area is $F = a \cdot f(a)$. Similarly, we now consider the triangle *RSL*, which in turn "frames" the fan curve. Here, the area is $G = T \cdot (g(T) + 1)/2$. We now use (6.1) and (6.3) for $x = a$ and $t = T$. This gives us

$$\begin{aligned} F = a \cdot f(a) &= \frac{T}{g(T)+1} \cdot f\left(\frac{T}{g(T)+1}\right) \\ &= \frac{T \cdot (g(T)+1)^2}{2(g(T)+1)} = \frac{T \cdot (g(T)+1)}{2} = G. \end{aligned}$$

This means that the areas of the rectangle and the triangle are equal. Thus, we not only know that the original integral is equal to the area under the fan curve (white and green), but also that the remaining areas (blue and brown) are equal. These remaining areas belong to the inverse functions of f and g. This gives us the opportunity to find new connections and new formulas.

6.5 The Power Function

We will now show more applications of Theorem 1. For $f(x) = \frac{x^k}{2}$ we get the following equation with (6.3) $t^k = (g(t) + 1)^{k+2}$ or $g(t) + 1 = t^{\frac{k}{k+2}}$. For $k = 1, 2$ the exponent in the fan curve is a unit fraction and the inverse function is a power function with a natural exponent. But already for $k \geq 3$ we no longer get unit

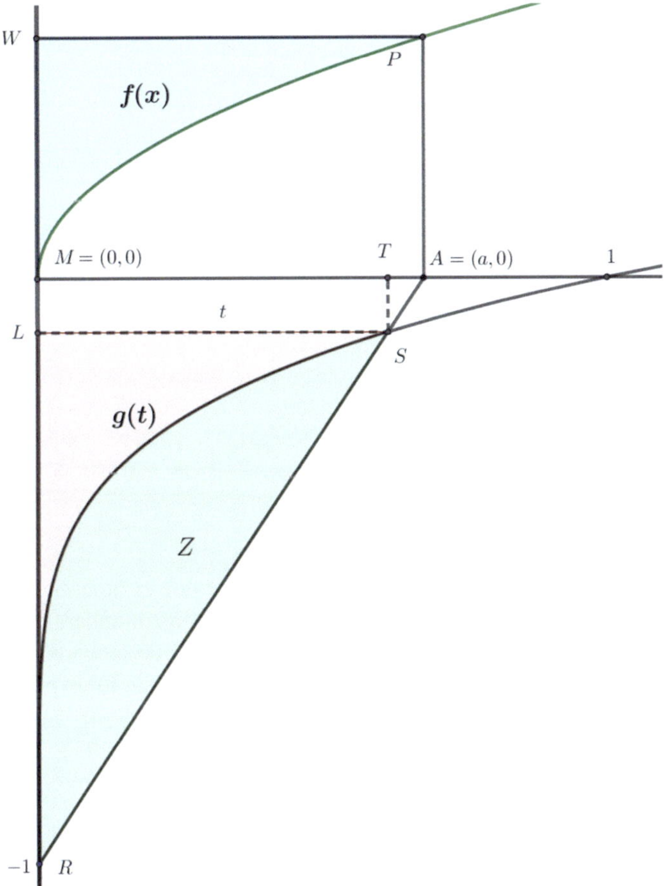

Fig. 6.5 Residual areas

fractions. It is therefore worth setting $f(x) = \frac{x^{1/n}}{2}$. Because of (6.3), the fan curve is given by $t = (g(t) + 1)^{2n+1}$ or $g(t) = \sqrt[2n+1]{t} - 1$ and at least the inverse functions are power functions with natural exponents, so that we can use the above technique using inverse functions. The inverse function of $y = \frac{x^{1/n}}{2}$ is $(2x)^n = y$ and for $g(t) = \sqrt[2n+1]{t} - 1$ the inverse function is $(t+1)^{2n+1} = g(t)$. Here we shift the integration interval from $\left[-1, g(T)\right]$ to $\left[0, g(T) + 1\right]$ and we can then use the function t^{2n+1}. The equality of the remaining areas reads as follows:

$$\int_0^{f(a)} (2x)^n dx = \int_0^{g(T)+1} t^{2n+1} dt.$$

Because of $f(a) = f\left(\frac{T}{g(T)+1}\right) = \frac{1}{2}(g(T)+1)^2$, we can rewrite the last equation and get

$$\int_0^{\frac{(g(T)+1)^2}{2}} (2x)^n dx = \int_0^{g(T)+1} t^{2n+1} dt \text{ or } 2^n \int_0^{\frac{V^2}{2}} x^n dx = \int_0^V x^{2n+1} dx, \tag{6.4}$$

where $V = g(T) + 1$. With this, we have achieved a result that provides us with many "new" integration formulas. Starting from $f(x) = x$ the method delivers the integrals for $x^3, x^7, x^{15}, x^{31}, x^{63}, \ldots$ and starting from $f(x) = x^2$ the integrals for $x^5, x^{11}, x^{23}, x^{47}, \ldots$. For example, we have

$$\int_0^V x^5 dx = 4 \int_0^{V^2/2} x^2 dx = 4\left[\frac{x^3}{3}\right]_0^{V^2/2} = 4 \cdot \frac{1}{3} \cdot \frac{V^6}{8} = \frac{V^6}{6},$$

a result that could also be achieved by substitution.

So far, we have assumed that the exponents of our power functions are positive numbers. But now we are tempted to also allow negative numbers as exponents. However, as soon as negative exponents occur, we must be prepared for the integrals that start at zero to be improper integrals or even infinite for exponents ≤ -1. We can avoid this problem by starting the integration interval at 1 instead of 0.

6.6 The Hyperbola

The case $n = -1$ is particularly interesting because the integral here is not trivial and because the values of the exponents in formula (6.4) match ($n = 2n + 1$). This corresponds to $f(x) = 1/(2x)$ and $g(t) + 1 = 1/t$ at the beginning of Sect. 6.5. But because the coefficients in both cases are different, we can hope that the equality of areas actually brings us an interesting result about the hyperbolic integral. Therefore, we now study the integral of the function $y = \frac{1}{2x}$ (see Fig. 6.6). With the same arguments as above and using $A = (a, 0)$, we can show that the area under the hyperbola $y = 1/(2x)$ over the interval $[1, a]$ is equal to the area of the curvilinearly bounded "triangle" RQW. Now, the triangle REW has the area $1/2$. The area of the triangle RSQ is exactly the same, because $\frac{1}{2}T \cdot \frac{1}{T} = \frac{1}{2}$, which means that the triangle RUW and the trapezoid $ESQU$ are also of the same size. Therefore, we have

$$\int_1^a \frac{dx}{2x} = \int_1^T \frac{dx}{x} \text{ and since } a = \frac{T}{g(T)+1} = \frac{T}{\frac{1}{T}} = T^2 \text{ this gives us}$$

$$\int_1^{T^2} \frac{dx}{x} = 2\int_1^T \frac{dx}{x}.$$

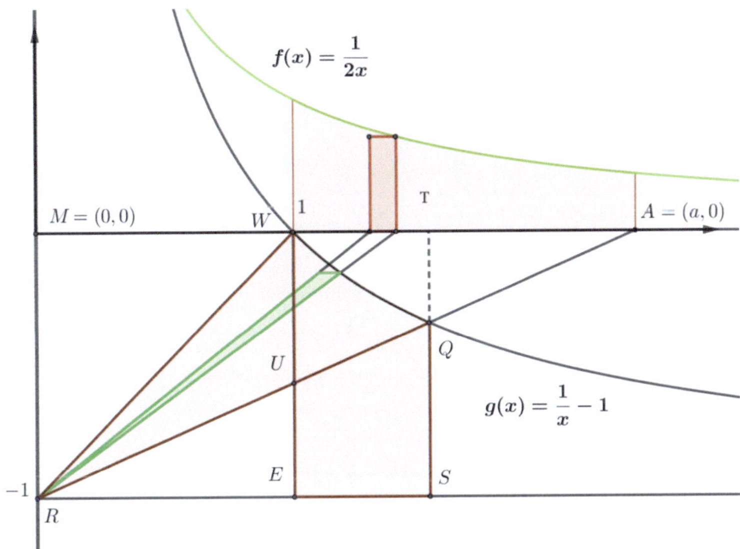

Fig. 6.6 The integration of the hyperbola

This is consistent with the previous result $2^n \int_0^{V^2/2} x^n dx = \int\limits_0^V x^{2n+1} dx$, where $n = -1$ and the lower limit of the integral is 1 instead of 0. If we now define $L(t) = \int_1^t \frac{dx}{x}$, we get $L(T^2) = 2 \cdot L(T)$, which is precisely the logarithmic property of the hyperbolic integral.

6.7 The Function $f(x) = \frac{1}{2x^2}$

Using formula (6.3) we get $\frac{1}{\frac{2t^2}{(g(t)+1)^2}} = \frac{(g(t)+1)^2}{2}$ or $t^2 = 1$, which gives us $t = \pm 1$. We choose $t = 1$ to correspond to the right part of the hyperbola. The similarity of the triangles gives us $\frac{a}{1} = \frac{1}{K}$. Therefore, the area of the triangle RFB is equal to $\frac{1-K}{2} = \frac{1}{2} - \frac{1}{2a}$. So, we have $\int_1^a \frac{dx}{2x^2} = \frac{1}{2} - \frac{1}{2a}$. Due to the vertical lines, the situation here is particularly simple. At the same time, we see that the fan curve g, which is given here by $t = 1$, is no longer a function (see Fig. 6.7). On the other hand, we see that some curvilinearly bounded areas can be easily represented by area-equivalent linearly bounded triangle areas.

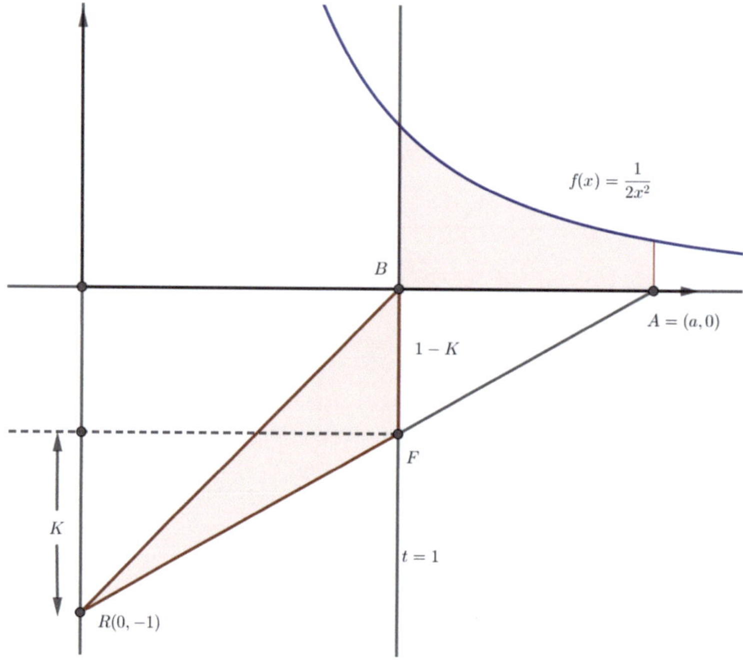

Fig. 6.7 Integration of a higher hyperbola

6.8 The "Nose Curves"

Our next example is a whole class of functions

$$f(x) = \frac{1}{2(x+1)^n}.$$

Equation (6.3) gives us $\frac{1}{2\left(1+\frac{t}{1+g(t)}\right)^n} = \frac{1}{2}(g(t)+1)^2$, thus

$$(1 + g(t))^{n-2} = (1 + g(t) + t)^n.$$

This looks like a "nose curve". In Fig. 6.8 we see the nose curve for $n = 3$. Here, the improper integral is $\int_0^\infty \frac{dx}{2(x+1)^3} = 1/4$, and therefore the "nose area" must be exactly $\frac{1}{4}$. This is again an example where the fan curve g is no longer a function.

Remark 6.4
The case $n = 2$ in Fig. 6.9 is particularly simple, because here we have

$$1 = (1 + g(t) + t)^2.$$

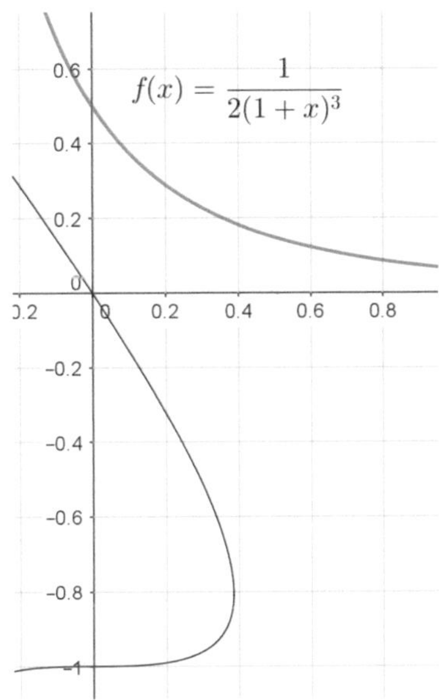

Fig. 6.8 The "nose curve"

Here, there are again two solutions. We choose $g(t) = -t$, because the second case $g(t) = -2-$t lies below the line $y = -1$. For linear functions, we can determine the integral with simple geometric tools. Since $A = (a, 0)$ it can be seen that $\int_0^a \frac{dx}{(1+x)^2}$ must be equal to the area of the triangle RZM. This is calculated to be $\frac{1}{2} \cdot 1 \cdot \frac{a}{a+1}$, because the x-value of Z is exactly $\frac{a}{a+1}$. This example can therefore be used to show that

$$\int_0^a \frac{dx}{(1+x)^2} = 1 - \frac{1}{1+a}.$$

Thus, we have succeeded in demonstrating the corresponding result from integral calculus with simple geometric means. Furthermore, it is immediately apparent that $\int_0^\infty \frac{dx}{2(1+x)^2}$ must be finite, because the intersection point Z never falls below the line $y = -1$. Therefore, the integral must always be less than or equal to $\frac{1}{2}$. We used a similar argument in our first example by Aage Bondesen to show that $\int_0^\infty \frac{dx}{2(1+x^2)} = \frac{\pi}{4}$.

Remark 6.5
The case $n = 1$ can give us insight in another way. If we choose $f(x) = \frac{1}{2(x+1)}$, Eq. (6.3) gives us $\frac{1}{2\left(\frac{t}{g+1}+1\right)} = \frac{1}{2}(g+1)^2$. This is again equivalent to $\frac{g(t)+1}{2(t+g(t)+1)} = \frac{1}{2}(g(t)+1)^2$. From this we get $(t + g(t) + 1)(g(t) + 1) = 1$ or $(g(t) + 1)^2 + t(g(t) + 1) - 1 = 0$. According to the p, q formula, this results in

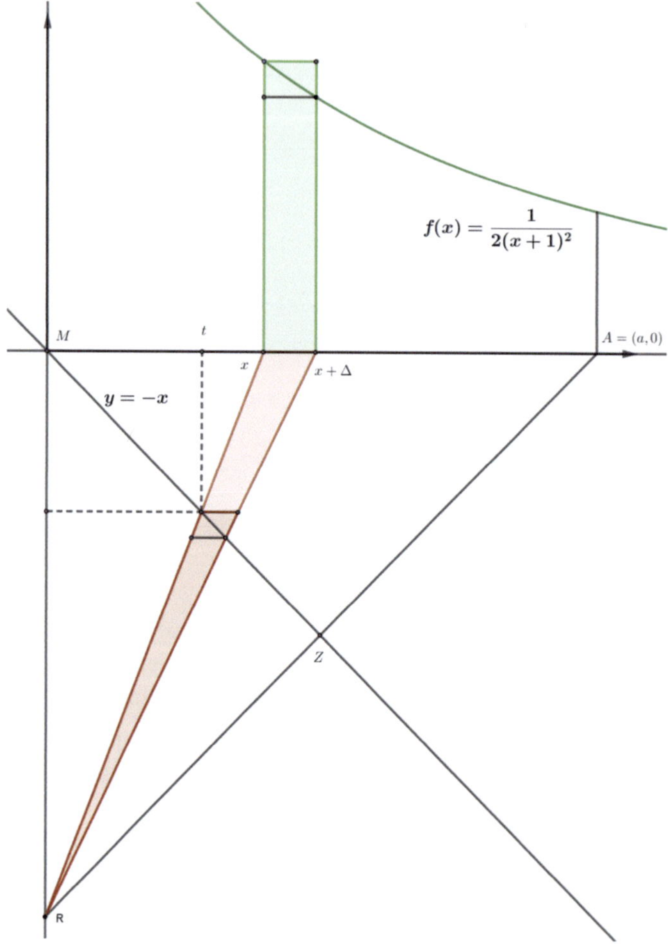

Fig. 6.9 Integration of another higher hyperbola

$g(t) + 1 = \frac{-t \pm \sqrt{4+t^2}}{2}$. Thus, the integral $\int \frac{1}{2(x+1)} dx$, which we know to be a logarithmic function, can be related to the integral $\int \left(\frac{-t \pm \sqrt{4+t^2}}{2} \right) dt$, a fact that allows us to determine the integral $\int \sqrt{a^2 + t^2} dt$. This explains why logarithms appear in such integrals. The details are complicated and are omitted here.

6.9 The Bell Curves

The following example is due to Stephan Berendonk (University of Wuppertal). We consider a variant of the original bell curve, the Witch of Agnesi, in our first example. In our new example, we replace the Witch of Agnesi with the function $f(x) = \frac{1}{2(1+x^2)^2}$. Because of Eq. (6.3) we have

$$\frac{1}{\left(1+\left(\frac{t}{g(t)+1}\right)^2\right)^2} = (g(t)+1)^2 \ \text{ or } \ g(t)+1 = (g(t)+1)^2 + t^2,$$

which represents a circle with center $\left(0, -\frac{1}{2}\right)$ and radius ½ (see Fig. 6.10). This is also another example where the fan curve no longer represents a function. Strictly speaking, the above equation gives us a pair of circles. However, we will ignore the lower circle since it lies below the line $y = -1$. Since the area of the semicircle is $\frac{1}{2}\pi\left(\frac{1}{2}\right)^2 = \frac{\pi}{8}$, we can deduce that

$$\int_0^\infty \frac{dx}{2\left(1+x^2\right)^2} = \frac{\pi}{8}, \ \text{ or } \int_0^\infty \frac{dx}{\left(1+x^2\right)^2} = \frac{\pi}{4},$$

a result that can be found in many integration tables.

In fact, the entire class of bell curves

$$f(x) = \frac{1}{2\left(1+x^2\right)^n},$$

which includes both the Witch of Agnesi ($n = 1$) and the curve of Berendonk ($n = 2$), can be handled with our method. For higher values of n we use Eq. (6.3) and get

$$\frac{1}{\left(1+\left(\frac{t}{g+1}\right)^2\right)^n} = (g+1)^2 \ \text{ or } \ (g+1)^{2(n-1)} = \left((g+1)^2 + t^2\right)^n$$

which represents a pair of ovals.

For $n = 1$ we get a single circle. For larger values of n terms of the form $(g+1)^2$ appear on both sides, so that pairs of ovals are formed (see Fig. 6.11 for the case $n = 3$). For our purposes, however, only the upper branches above $y = -1$ are of interest.

We now want to determine the area of the ovals and take advantage of the fact that the integral of $f(x)$ from zero to infinity can actually be calculated. We set

$$I_n = \int_0^\infty \frac{dx}{\left(1+x^2\right)^n}.$$

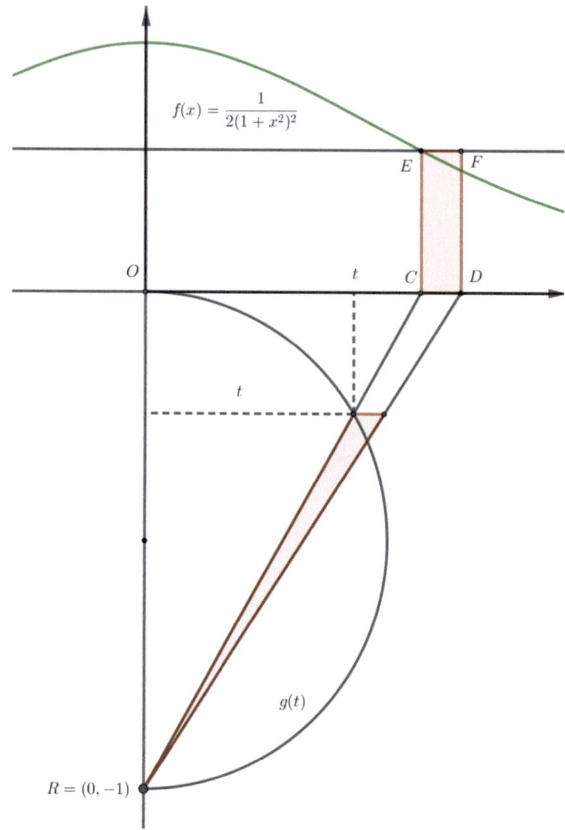

Fig. 6.10 Integration of the bell curve for $n = 2$

Then we have $I_n = \int_0^\infty \frac{dx}{(1+x^2)^n} = \int_0^\infty \frac{1+x^2-x^2}{(1+x^2)^n} dx = I_{n-1} - \int_0^\infty \frac{x \cdot x dx}{(1+x^2)^n} = I_{n-1} - \int_0^\infty u \cdot v' dx$, where u and v are defined as follows: $u = x$ and $v = \frac{-1}{2(n-1)} \cdot \frac{1}{(1+x^2)^{n-1}}$. We now perform integration by parts and get

$$\int_0^\infty u \cdot v' dx = u \cdot v|_0^\infty - \int_0^\infty u' \cdot v dx = 0 - \left(\frac{-1}{2(n-1)}\right) \int_0^\infty \frac{dx}{\left(1+x^2\right)^{n-1}}.$$

All in all, this gives us

$$I_n = I_{n-1} - \left(\frac{1}{2(n-1)}\right) I_{n-1} = \left(\frac{2n-3}{2n-2}\right) I_{n-1},$$

which then brings us the following

$$I_n = \left(\frac{2n-3}{2n-2}\right)\left(\frac{2n-5}{2n-4}\right) \cdots \frac{3}{4} \cdot \frac{1}{2} \cdot \frac{\pi}{2},$$

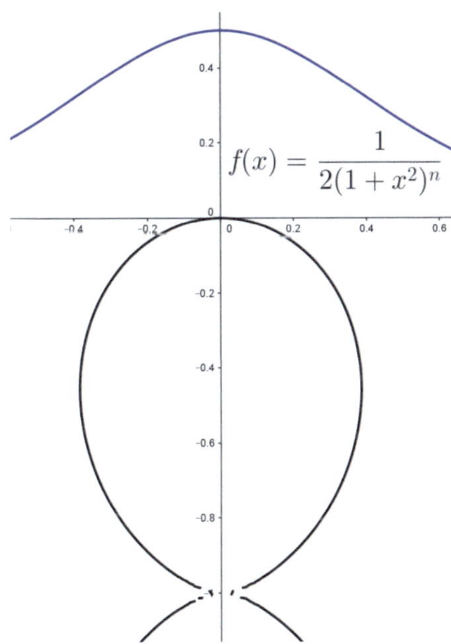

Fig. 6.11 A pair of ovals that belong to the bell curve

because $I_1 = \frac{\pi}{2}$, as we saw in the first example with the Witch of Agnesi. If we multiply this result by 2, we get the area of the ovals.

6.10 Another Parabola

What about the parabola $f(x) = 1 - x^2$? Here Eq. (6.3) gives us the following

$$1 - \left(\frac{t}{g(t)+1}\right)^2 = \frac{1}{2}(g(t)+1)^2 \text{or}$$
$$2(g(t)+1)^2 - 2t^2 = (g(t)+1)^4.$$

This is the equation of a lemniscate. Now, we know the area under the original curve, the parabola, over the interval [0, 1]. We have $\int_0^1 (1-x^2)dx = 2/3$. Thus we also know the area of the lemniscate. It is $4 \cdot \frac{2}{3} = \frac{8}{3}$ (see Fig. 6.12). In this example, we no longer used the factor 2 in the denominator. The effect is then that $g(0)$ can be greater than zero, which is also the case in Figs. 6.12, 6.13, and 6.14. This is not a big problem, as the original goal was only to keep the graph of g below the x-axis to better visualize the method.

Remark 6.6
By changing the shape of the parabola $y = a^2 - b^2x^2$ we can generate many different forms of lemniscates.

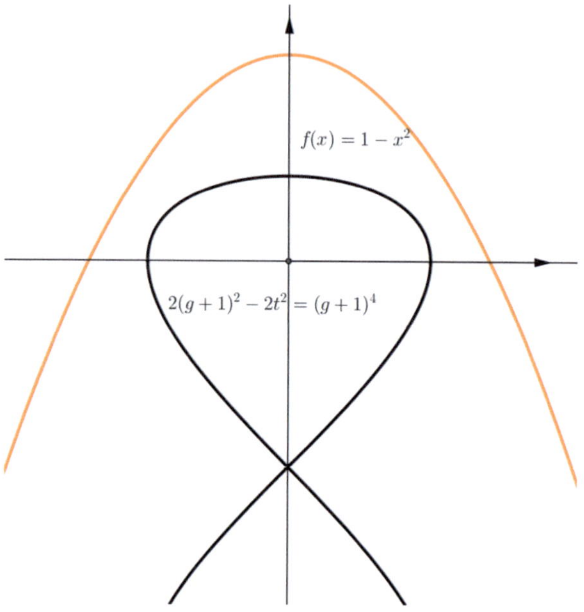

Fig. 6.12 A lemniscate as the fan curve of a parabola

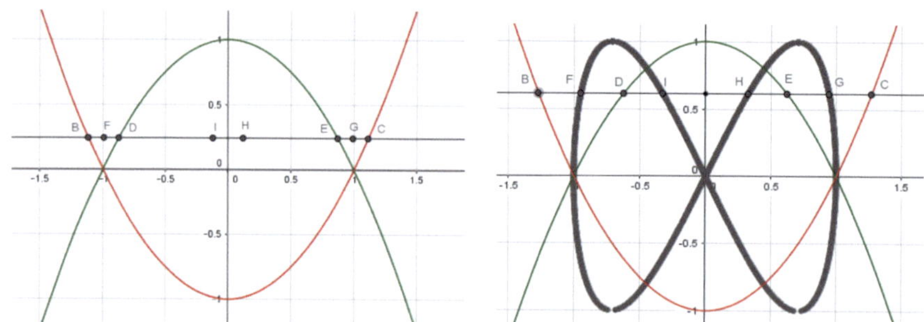

Fig. 6.13 The lemniscate of Gerono

Remark 6.7

The lemniscate $2(g+1)^2 - 2t^2 = (g+1)^4$ is known as the lemniscate of Gerono. It was already described by Gregorius of St. Vincent. However, it was named after the French mathematician Camille-Christophe Gerono (1799–1891), see Haftendorn (2017, p. 117) and Lawrence (2014, p. 124). It is defined as a so-called center curve: We start with two parabolas e.g., $y = 1 - x^2$ and $y = x^2 - 1$ with the same vertical line of symmetry. One curves upwards, the other downwards. Now we cut the pair of parabolas with horizontal lines (BC in Fig. 6.13) and mark the intersection points (B, D, E and C). The midpoints (F, I, H and G) between the intersection points of

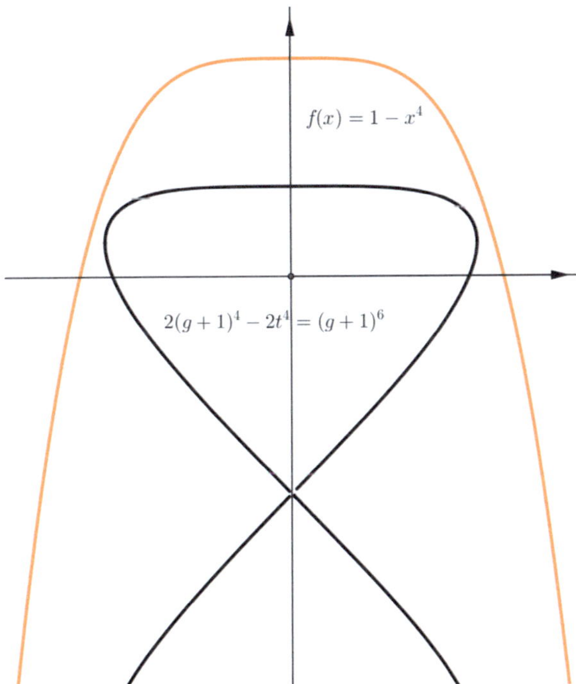

Fig. 6.14 Lemniscate for $n = 2$

one parabola with the horizontal lines and those of the other parabola with the same horizontal lines make up the points of the lemniscate of Gerono, when we let the horizontal lines (BC) vary.

Remark 6.8
It can be shown that parabolas with a maximum point, which have roots to the right and left of the origin, always give us lemniscates. Parabolas that have a minimum point give us other curves. The root function from Sect. 6.4 is evidence of this. In general, the fan curve of a parabola is a curve of degree four.

The entire class of functions $f(x) = 1 - x^{2n}$ can be treated in exactly the same way as the parabola from Sect. 6.10. Here too, lemniscate-like figures are created. But since $\int_0^1 (1 - x^{2n})dx = \frac{2n}{2n+1}$, the area of these lemniscate-like figures must be $\frac{8n}{2n+1}$.

In Fig. 6.14 we see the graph of the function $f(x) = 1 - x^4$ and the associated lemniscate $2(g+1)^4 - 2t^2 = (g+1)^6$ with a total area of $\frac{16}{5}$.

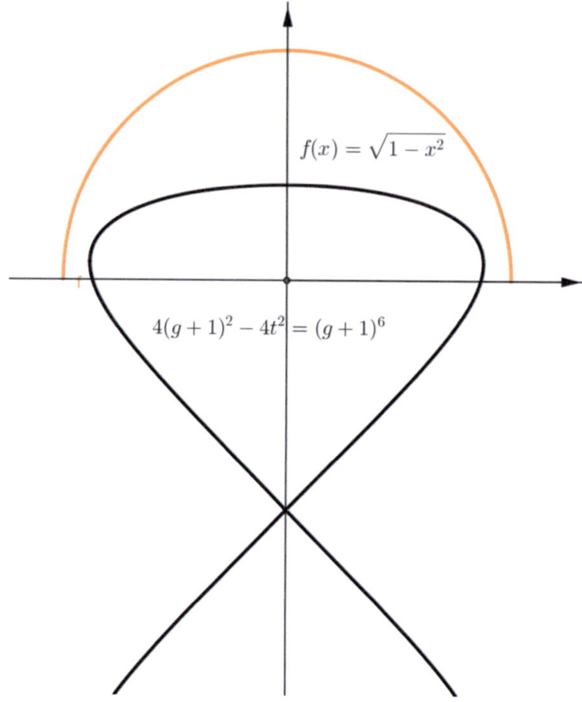

Fig. 6.15 A semicircle and the corresponding lemniscate

6.11 The Semicircle

If we choose a semicircle $f(x) = \sqrt{1-x^2}$ for $-1 \leq x \leq 1$, then due to (6.3) the fan curve is

$$4\big((g(t)+1)^2 - t^2\big) = (g(t)+1)^6.$$

This also corresponds to a lemniscate. Because this curve originates from a semicircle, the area of the entire lemniscate is $2 \cdot \frac{\pi}{2} = \pi$ (see Fig. 6.15).

6.12 The Exponential Function

If we choose the exponential function as the initial function $f(x) = \frac{e^x}{2}$, we obtain the following relationship with Eq. (6.3) $e^{\frac{t}{g+1}} = (g+1)^2$ and thus

$$e^t = (g+1)^{2(g+1)} = e^{2(g+1)\ln(g+1)}.$$

Since both sides of the equation are powers of e, the exponents must also be equal and we obtain for the fan curve

$$t = 2(g+1)\ln(g+1).$$

With the fan method, we can now determine the area under this function. The inverse function of this function is $y = 2(x+1)\ln(x+1)$. We will instead work with the mirror curve because we can more easily recognize it as a function of the variable x, namely $y = 2(x+1)\ln(x+1)$. We can now expect that we can also calculate the integral of this inverse curve with the fan method. This will make it easy for us to also integrate $y = x\ln(x)$. We assume the integral of the exponential function as known.

Here we see that $\frac{1}{2}\int_0^a e^x dx = Z$, where Z is the green area between the y-axis, the fan curve $x = 2(y+1)\ln(y+1)$ and the straight line that connects $R = (0,-1)$ and $A = (a,0)$ (see Fig. 6.16). We now reflect Z and the mentioned fan curve about line $y = x$ and obtain the new area Z' with the same area and the new curve $y = 2(x+1)\ln(x+1)$. The mirror image of the straight line that connects R and $A = (a,0)$, is the straight line $y = a(x+1)$. This goes through the mirror points $R' = (-1,0)$ and $A' = (0,a)$. The area equality, which is guaranteed to us by the fan method, then yields

$$\begin{aligned}\frac{1}{2}\int_0^a e^x dx &= \frac{e^a-1}{2} = Z = Z' \\ &= \int_0^k (a(x+1) - 2(x+1)\ln(x+1))dx + \frac{a}{2} \\ &= \int_1^{k+1} (ax - 2x\ln(x))dx + \frac{a}{2}.\end{aligned}$$

The last expression $\frac{a}{2}$ stands for the area of the triangle $R'OA'$. This leads us to

$$2\int_1^{k+1} x\ln(x)dx = \int_1^{k+1} ax\,dx + \frac{a}{2} - \frac{e^a-1}{2} = \frac{a(k+1)^2}{2} - \frac{e^a-1}{2}.$$

Here, k is the x-value of the intersection point of the straight line $y = a(x+1)$ and the mirrored fan curve $y = 2(x+1)\ln(x+1)$ (see Fig. 6.17). So, we have

$$a(k+1) = 2(k+1)\ln(k+1) \text{ or } a = 2\ln(k+1).$$

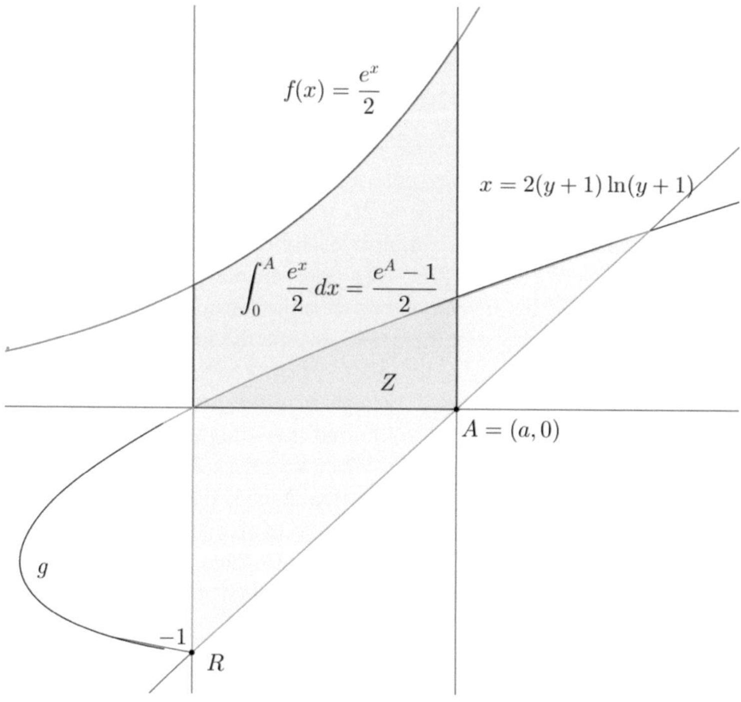

Fig. 6.16 The fan curve of the exponential function

We can now substitute this into the integration formula and get

$$2\int_{1}^{k+1} x\ln(x)dx = (k+1)^2\ln(k+1) - \frac{(k+1)^2 - 1}{2}.$$

This then implies

$$\int_{1}^{q} x\ln(x)dx == \frac{q^2\ln(q)}{2} - \frac{q^2}{4} + \frac{1}{4}.$$

The last term $\frac{1}{4}$ represents the area under the curve between 0 and 1. Thus, we have found $\int x\ln(x)dx$, which was our goal here.

We now continue by generalizing our results again. So far, we have only considered functions above the x-axis. However, it turns out that the relationships become simpler when we move from functions to parametric curves. Indeed, the transition Eq. (6.3) from f to g becomes much simpler when we use parametric curves instead of functions.

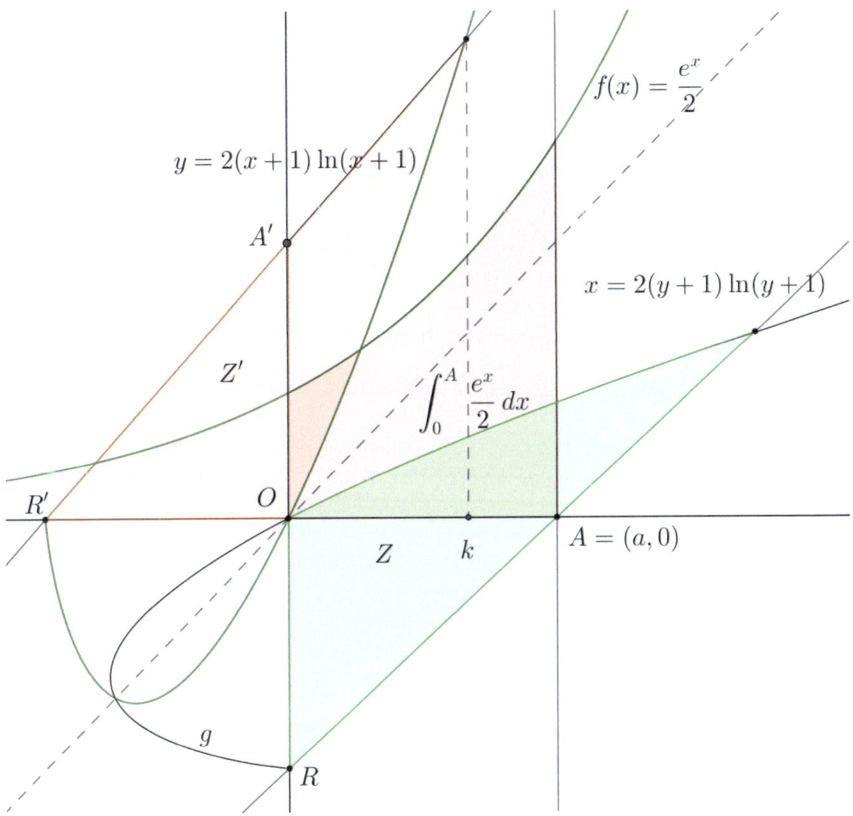

Fig. 6.17 Mirrored fan curve

6.13 Parametric Representations

Given is a continuous curve above the x-axis through a parametric representation $\underline{x} = x(s), \underline{y} = y(s)$. Now we look at the associated fan curve $x = t(s), y = g(s)$. The area of a Riemann rectangle is $y(s) \cdot \Delta$, while the associated triangle has the area $\frac{\Delta}{2} \cdot \left(\frac{t(s)}{x(s)}\right)^2$. But since both should be equal, we have

$$y(s) = \frac{1}{2}\left(\frac{t(s)}{x(s)}\right)^2, \quad \text{thus} \quad t(s) = x(s) \cdot \sqrt{2y(s)}.$$

In addition, we have the equation that describes the similarity

$$\frac{x(s)}{1} = \frac{t(s)}{g(s)+1} \quad \text{or} \quad g(s)+1 = \frac{t(s)}{x(s)} = \frac{x(s)\sqrt{2y(s)}}{x(s)} = \sqrt{2y(s)}.$$

Thus, we have found a very simple relationship between the parametric representation of the initial curve above the x-axis and the parametric representation of the fan curve below the x-axis. We would like to formulate this result as a theorem.

Theorem 2 Given is a continuous parametric curve $\underline{x} = x(s), \underline{y} = y(s)$. The associated fan curve then has the parametric representation:

$$\underline{x} = t(s) = x(s) \cdot \sqrt{2y(s)} \text{ and } \underline{y} + 1 = g(s) + 1 = \sqrt{2y(s)}.$$

Remark 6.9
If the initial curve above the x-axis is already a function, we have $\underline{x} = x$ and $\underline{y} = f(x)$. Then $g + 1 = \sqrt{2f(x)}$, which gives us $\frac{(g+1)^2}{2} = f(x) = f\left(\frac{t}{g+1}\right)$ and we have proven Theorem 1 once again.

Remark 6.10
With the same formula, it is also possible to reconstruct the original initial curve $(x(s), y(s))$ if the fan curve $(t(s), g(s))$ is given as a parametric curve. We have namely $y(s) = \frac{(1+g(s))^2}{2}$ and $x(s) = \frac{t(s)}{g(s)+1}$. However, we do not necessarily obtain the initial curve as a function $\underline{y}$ of $\underline{x}$ but again as a parametric curve.

Remark 6.11
In the introduction to this chapter, we mentioned that Fermat (1896) was the first to integrate the Witch of Agnesi and that he used two variable transformations in the process. For him, the Witch of Agnesi was defined by the equation.

$$b^3 = xy^2 + b^2x.$$

Compared to our representation, the variables x and y are exchanged. The quantity b here is a constant. In our case, $b = 1$. As mentioned, Fermat carries out two variable transformations (see Jaume Paradis et al. 2008). In the first, he sets $x = z^2/b$, which corresponds to $y = (g + 1)^2/2$ in our case, if we disregard the factor ½. In the second, Fermat sets $y = bu/z$, which then corresponds to $x = t/(g + 1)$ in our case. By shifting from g to $g + 1$ their curve becomes our fan curve. This was necessary only for technical reasons in our case, because we wanted to visually separate the corresponding illustrations. Thus, we see that Fermat's variable transformations largely coincide with the transformations here, if we use the parametric version of the method from Theorem 2. In this way, we can now give a geometric interpretation of Fermat's results, which were purely algebraic, and we can identify Fermat's method with the method of Aage Bondesen. Jaume Paradís et al. (2008, p. 5), who study Fermat's *Treatise*, write: "This second part of the *Treatise* is obscure and difficult to read." The Fermat method, which is opaque for many, has finally been given a geometric interpretation with the fan method presented here.

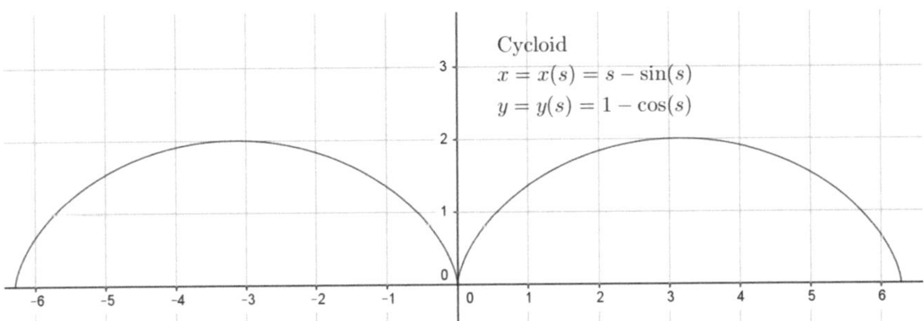

Fig. 6.18 The cycloid

Remark 6.12

If we use the parametric representation of the unit circle $x(s) = \cos(s)$ and $y(s) = \sin(s)$ from Sect. 6.11, we get $x = t(s) = \cos(s)\sqrt{2\sin(s)}$ and $y + 1 = g(s) + 1 = \sqrt{2\sin(s)}$. This gives us the same lemniscate $4t^2 = 4(g+1)^2 - (g+1)^6$, which we had already found in Sect. 6.11 (see also Fig. 6.14).

6.14 The Cycloid

Now, we would like to apply our new findings. A classic curve that is most easily specified as a parametric curve is the cycloid.

The cycloid (Fig. 6.18) is the curve that is formed when we follow a point P on the periphery of a circle, which rolls on a horizontal line (x-axis). So, it describes, for example, the movement of a bicycle valve when the wheel rolls on a straight path.

Here $x(s) = s - \sin(s)$ and $y(s) = 1 - \cos(s)$ (see Fig. 6.19). This gives us $t(s) = (s - \sin(s))\sqrt{2(1 - \cos(s)}$ and $g(s) + 1 = \sqrt{2(1 - \cos(s))}$. This curve can be seen in Fig. 6.20. This fan curve, which originates from a cycloid, thus looks like the wing of a dragonfly, see Fig. 6.20. But since it is well known that a segment of a cycloid encloses an area of 3π, we can conclude that a pair of dragonfly wings covers a total area of 6π.

6.15 Surfaces in Space

Finally, we try to understand the method of Aage Bondesens in three dimensions. To do this, we first consider an example. From there, we then develop the general rules of play for the relationship between initial surfaces and corresponding surfaces, which we then again want to call “fan surfaces”. Then we consider new examples and calculate volumes of many previously “unknown” three-dimensional figures.

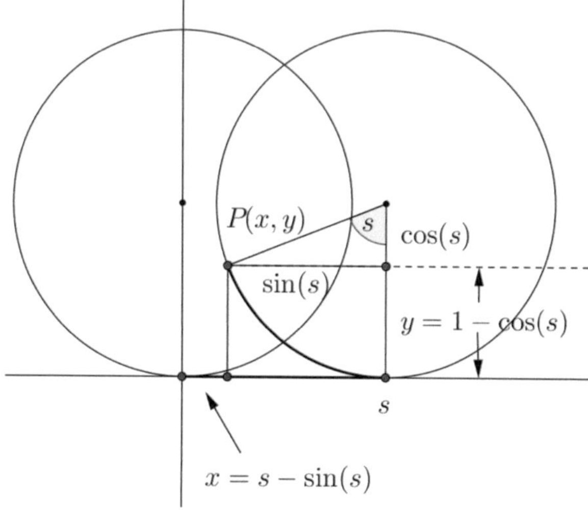

Fig. 6.19 The formula of the cycloid

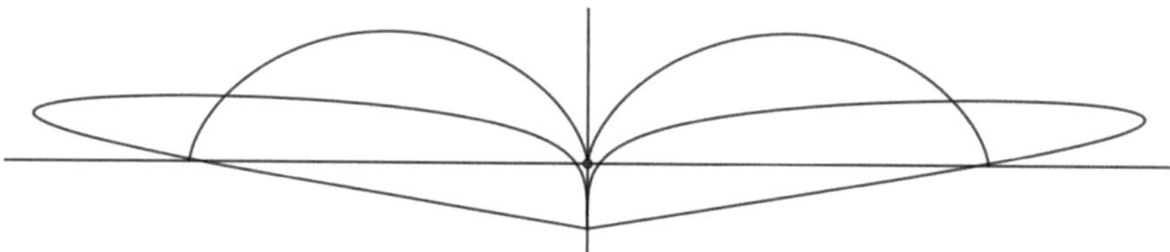

Fig. 6.20 Wings of a dragon fly

6.16 The Surface that Belongs to the Hemisphere Below the *xy*-Plane

We start with an example that clearly illustrates the procedure. We begin with the surface function

$$f(x, y) = \frac{1}{3\sqrt{1 + x^2 + y^2}^3}.$$

The graph can be seen in Fig. 6.21. Here we consider a "Riemann column" at the position (x, y) in the xy-plane. The base of the column then has the area Δ^2 (see Fig. 6.22). On the same base, we now construct a pyramid, with the vertex at $R = (0, 0, -1)$. This pyramid then has the volume $\Delta^2/3$, because the height is equal to 1. Now we reduce the pyramid until the reduced base area reaches a unit sphere, which has been shifted by one unit in the negative z-direction. During the reduction, we shrink the pyramid into itself. The reduction factor is $\frac{1}{\sqrt{1+x^2+y^2}}$. Now the volume of the reduced wedge is $\frac{\Delta^2}{3\sqrt{1+x^2+y^2}^3}$, because we have to use the third power of the reduction factor, if we want to find the volume of the reduced wedge. Length, width, and height of the wedge are reduced by the same factor. However, the reduced wedge has exactly the same volume as the corresponding Riemann column $\Delta^2 \cdot f(x, y)$. By summing up all the columns over the entire plane, we obtain

$$\iint_{xy\text{-plane}} \frac{dxdy}{3\sqrt{1+x^2+y^2}^3} = \frac{2\pi}{3},$$

which is exactly the volume of a hemisphere with a radius of 1.

We now try to formulate this result in such a way that it can be easily applied to new examples. The result is summarized in Theorem 3.

Theorem 3 Consider a surface defined by a continuous function $f(x, y)$. The associated fan surface is then given by

$$f\left(\frac{t}{g(t,s)+1}, \frac{s}{g(t,s)+1}\right) = \frac{1}{3}(g(t,s)+1)^3.$$

The volumes under the initial surface and the fan surface are equal.

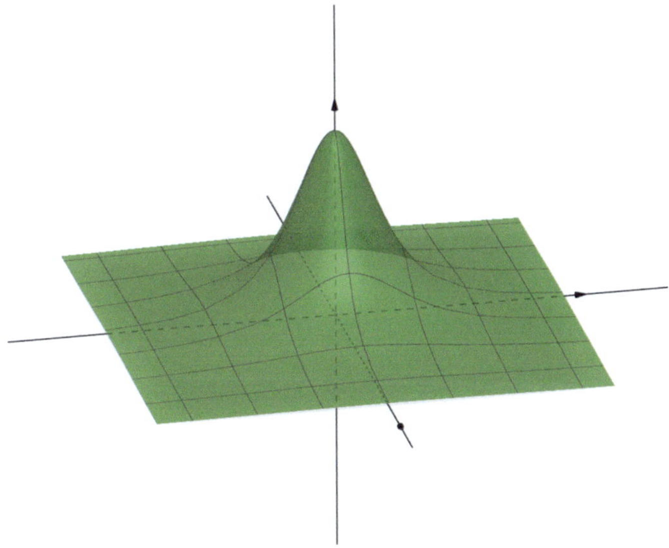

Fig. 6.21 Surface in space

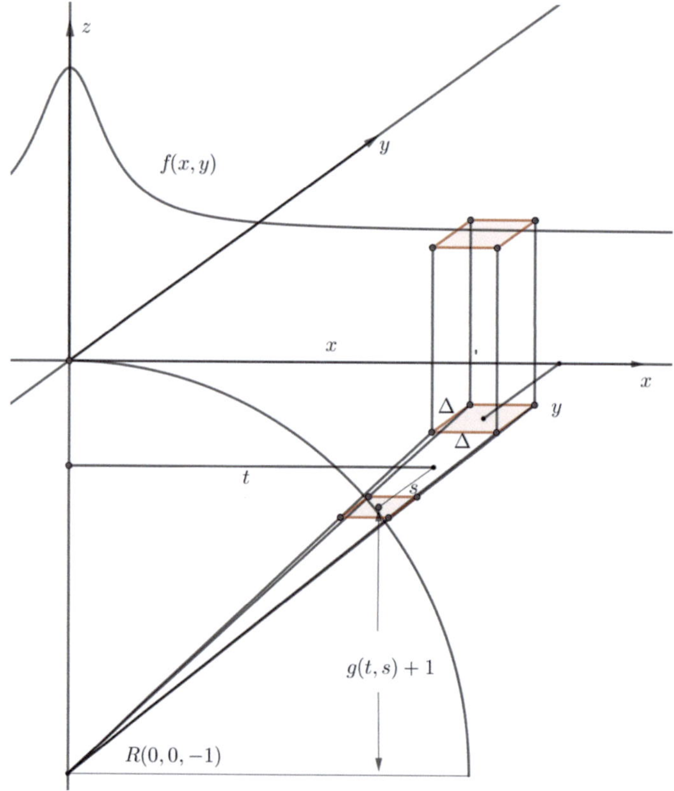

Fig. 6.22 A Riemann column is exchanged with a wedge

Proof. On the one hand, we have

$$\Delta^2 \cdot f(x, y) = \frac{\Delta^2}{3}\left(\frac{t}{x}\right)^3,$$

because the Riemann column and the pyramid, both built on the same base area Δ^2, are supposed to have the same volume. Here, $\frac{t}{x} = \frac{s}{y}$ is the reduction factor. The similarity gives us

$$\frac{x}{1} = \frac{t}{g(t, s) + 1} \quad \text{and} \quad \frac{y}{1} = \frac{s}{g(t, s) + 1}.$$

Thus, the fan surface satisfies the following equation, which strongly reminds us of (6.3)

$$f\left(\frac{t}{g(t, s) + 1}, \frac{s}{g(t, s) + 1}\right) = \frac{1}{3}(g(t, s) + 1)^3.$$

In the following, we would like to apply Theorem 3. We consider a series of known geometric solids, spheres, ellipsoids, cones and pyramids and calculate the associated fan surfaces. The new bodies that are created have the same volume as the bodies that we considered initially. This allows us to determine a number of previously "unknown" volumes. We can also proceed in reverse by starting with a known fan surface and we search for the associated initial function over the xy-plane. This gives us new results in some cases.

6.17 The Surface that Belongs to the Hemisphere Above the *xy*-Plane

In the previous example, the fan surface was a sphere. Now we turn the tables and choose a hemisphere as the initial surface above the xy-plane (see Fig. 6.23). A hemispherical surface has the following formula

$$x^2 + y^2 + z^2 = 1.$$

We consider here a variant of the hemispherical surface (Fig. 6.24) namely

$$9z^2 = 1 - x^2 - y^2 \text{ or } z = f(x, y) = \frac{\sqrt{1 - x^2 - y^2}}{3}.$$

This is a hemi-ellipsoid (see Fig. 6.24) and strongly resembles a hemisphere (see Fig. 6.23). Each z-value of the hemisphere is compressed to a third. The "sphere" now looks more like a lozenge shape. The volume is then also only a third of the associated hemisphere volume, namely $V = \frac{2\pi}{9}$.

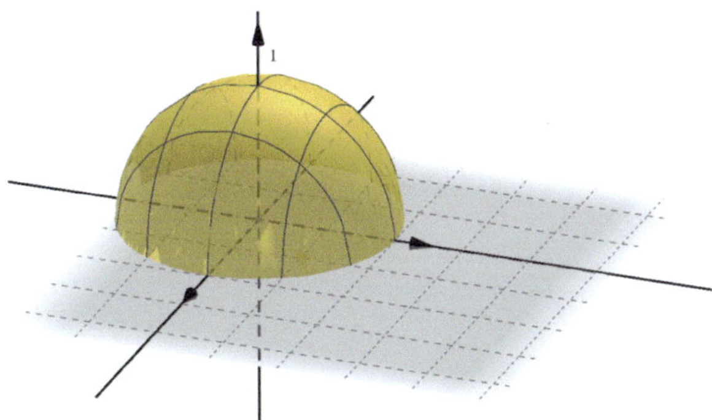

Fig. 6.23 Hemisphere

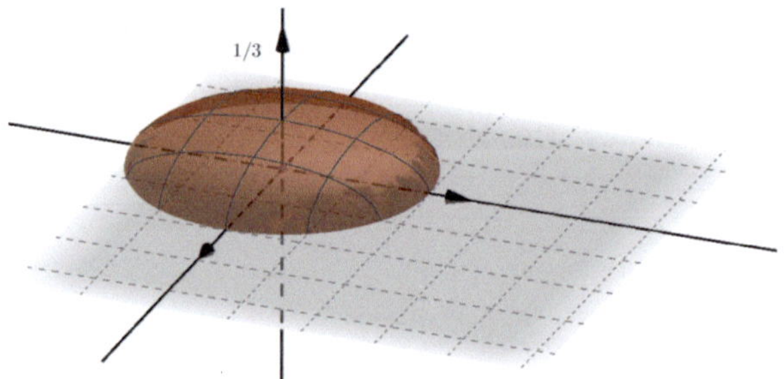

Fig. 6.24 Hemi-ellipsoid or lozenge

For the fan surface, we get here

$$\frac{1}{3}\sqrt{1-\left(\frac{t}{g+1}\right)^2-\left(\frac{s}{g+1}\right)^2}=\frac{(g+1)^3}{3}$$

or $1-\left(\frac{t}{g+1}\right)^2-\left(\frac{s}{g+1}\right)^2=(g+1)^6$, which in turn yields

$$(g+1)^2-t^2-s^2=(g+1)^8.$$

This surface represents a double gyroscope or a kind of diabolo (see Fig. 6.25), and the fan method guarantees us that the volumes of the diabolo and the ellipsoid are equal. Of course, we are only interested in the upper half of the diabolo, which lies above $z=-1$ and below the xy-plane. This "half diabolo" also has the volume $V=\frac{2\pi}{9}$. This reminds us of Sect. 6.11, where a circle was transformed into a lemniscate.

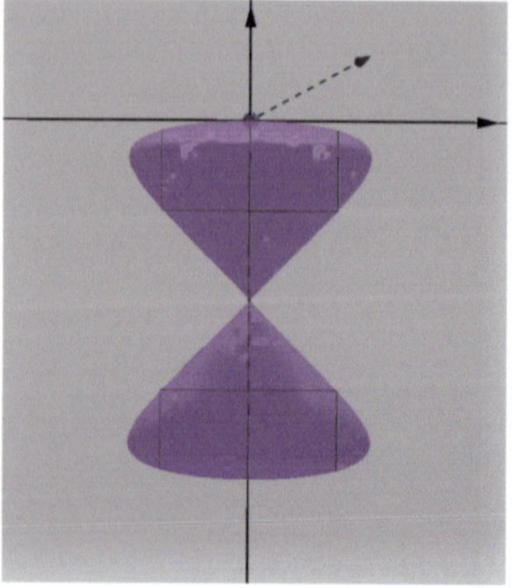

Fig. 6.25 A double gyroscope or diabolo

6.18 The Surface Belonging to a Cone Above the *xy*-Plane

Another well-known surface is the cone (see Fig. 6.26). Here we have

$$z = f(x, y) = \frac{1 - \sqrt{x^2 + y^2}}{3}.$$

The base is a unit circle with an area of π, while the height is 1/3. Thus, the volume is $V = \pi/9$.

Here we are only interested in the part of the cone that lies above the *xy*-plane. The associated fan surface then satisfies the equation

$$1 - \sqrt{\left(\frac{t}{g+1}\right)^2 + \left(\frac{s}{g+1}\right)^2} = (g+1)^3$$

or

$$(g+1)^2\left(1 - (g+1)^3\right)^2 = t^2 + s^2.$$

This equation describes a kind of onion (see Fig. 6.27). We only consider the part of this figure that lies between $z = 0$ and $z = -1$. Thanks to Theorem 3, we also know the volume of the onion. It has the same volume as the cone, namely $V = \pi/9$.

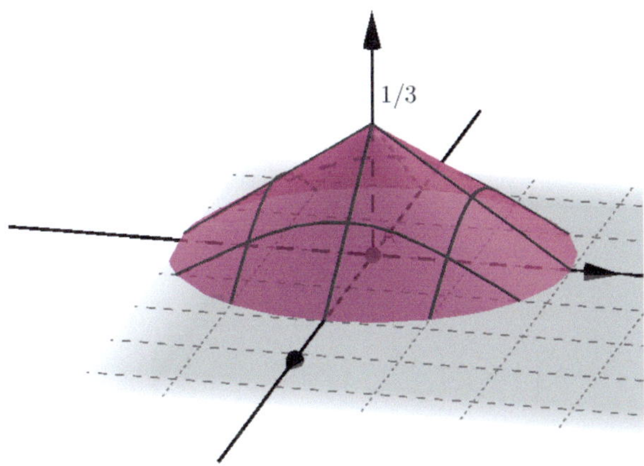

Fig. 6.26 Cone above the *xy*-plane

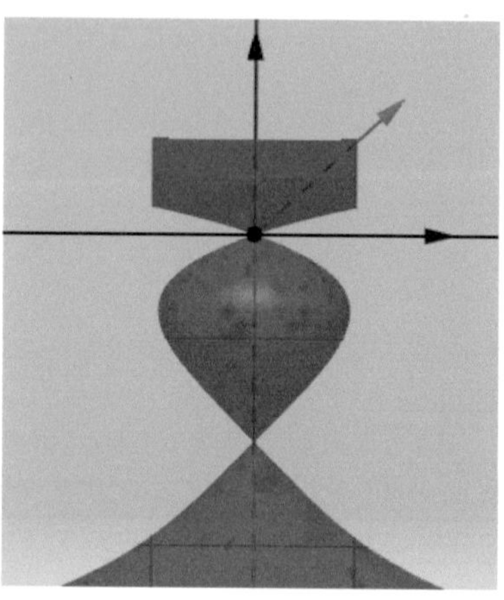

Fig. 6.27 Onion

6.19 Surface Belonging to a Cone Below the xy-Plane

Now we again interchange the roles and choose a cone as fan surface below the xy-plane and try to find the original starting surface $f(x, y)$. Here we have $g(t, s) = -\sqrt{t^2 + s^2}$.

We make a bold attempt, namely

$$f(x, y) = \frac{1}{3\left(1 + \sqrt{x^2 + y^2}\right)^3}.$$

Then we get

$$\frac{1}{3\left(1 + \sqrt{\left(\frac{t}{g+1}\right)^2 + \left(\frac{s}{g+1}\right)^2}\right)^3} = f\left(\frac{t}{g+1}, \frac{s}{g+1}\right) = \frac{(g+1)^3}{3},$$

so

$$1 = \left(1 + \sqrt{\left(\frac{t}{g+1}\right)^2 + \left(\frac{s}{g+1}\right)^2}\right)(g+1)$$

and thus

$$1 = g + 1 + \sqrt{t^2 + s^2}$$

and we actually get the above-described cone $g(t, s) = -\sqrt{t^2 + s^2}$ as a fan surface below the xy-plane. Let's now study the starting surface (see Fig. 6.28)

$$f(x, y) = \frac{1}{3\left(1 + \sqrt{x^2 + y^2}\right)^3}.$$

Above the xy-plane it looks like this:

The surface is unbounded with a maximum at $x = y = 0$ and we can read off that

$$\iint\limits_{xy\text{-plane}} \frac{dxdy}{3\left(1 + \sqrt{x^2 + y^2)}\right)^3} = \frac{\pi}{3},$$

because $\pi/3$ is just the volume of the cone, with which we started below the xy-plane. The height of the cone is one unit, while the base is a unit circle with an area of π. The corresponding starting function indeed has a sharp vertex at the origin. This can be easily seen, for example, when considering the intersection with the plane $y = 0$. There we have $z = f(x, 0) = \frac{1}{3(1+\sqrt{x^2+0^2})^3} = \frac{1}{3(1+|x|)^3}$. The absolute value function is known to have a sharp vertex at the origin (discontinuous derivative) and thus our suspicion is confirmed.

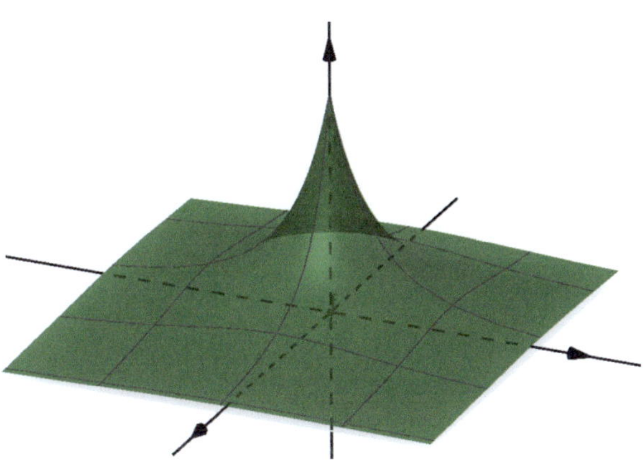

Fig. 6.28 Unbounded surface with a sharp vertex at the origin

6.20 Surface Belonging to a Paraboloid Above the xy-Plane

Also for a paraboloid (see Fig. 6.29) similar calculations can be made. Here we get

$$z = f(x, y) = \frac{1 - x^2 - y^2}{3}.$$

For the fan surface, we then have the following equation

$$1 - \left(\frac{t}{g+1}\right)^2 - \left(\frac{s}{g+1}\right)^2 = (g+1)^3$$

or

$$(1+g)^2 - t^2 - s^2 = (1+g)^5.$$

This time, the fan surface resembles a spindle (see Fig. 6.30). Now, we want to calculate the volume of the spindle. To do this, we need to calculate the volume of the associated paraboloid. We consider the paraboloid as a body of rotation, where the rotating function branch satisfies the equation $y = h(x) = \sqrt{x}$. Here, we imagine the paraboloid lying down to use the conventional method of volume calculation, as we know it from school (see Figs. 6.31 and 6.32).

Then we have

$$V = \pi \int_0^1 h^2(x)dx = \pi \int_0^1 x\,dx = \pi \left[\frac{x^2}{2}\right]_0^1 = \frac{\pi}{2}.$$

This is the volume of the paraboloid. With this, we also know the volume of the spindle, or the part of the fan figure between $z = 0$ and $z = -1$, namely $V = \frac{\pi}{2}$.

We will get to know the reverse case, where the paraboloid represents the fan surface in Sect. 6.26.

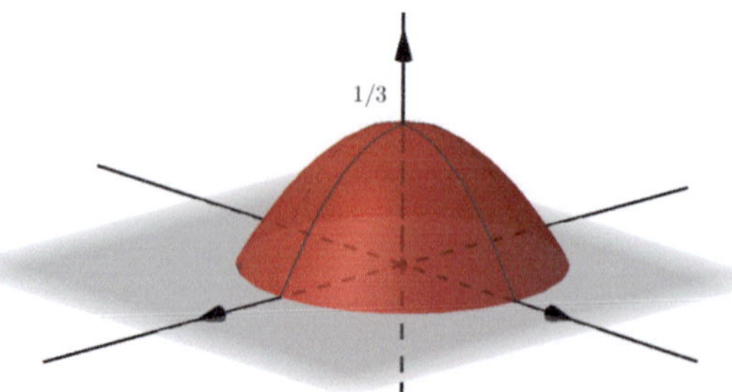

Fig. 6.29 Paraboloid

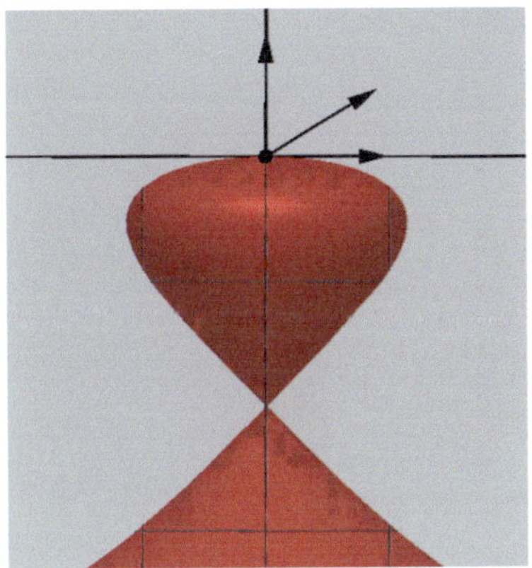

Fig. 6.30 Spindle

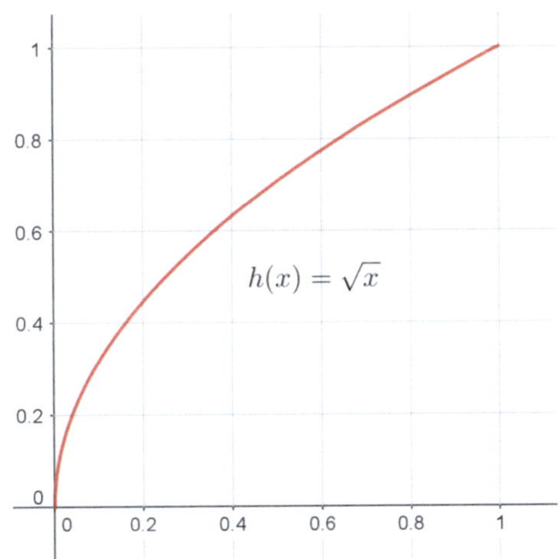

Fig. 6.31 Root function as starting point for the paraboloid

6.21 The Volume Under the Function $z = 1/3(1 + x + y)^3$

With the initial function $f(x, y) = \frac{1}{3(1+x+y)^3}$ we get a particularly beautiful and simple result. We consider the function only for positive values of x and y. With negative values, we risk the denominator becoming zero, which we definitely want to avoid.

In Fig. 6.33, we have limited ourselves to positive x- and y-values. Now, if we calculate the fan surface, the following happens:

$$\frac{1}{3\left(1 + \frac{t}{g+1} + \frac{s}{g+1}\right)^3} = f\left(\frac{t}{g+1}, \frac{s}{g+1}\right) = \frac{(g+1)^3}{3},$$

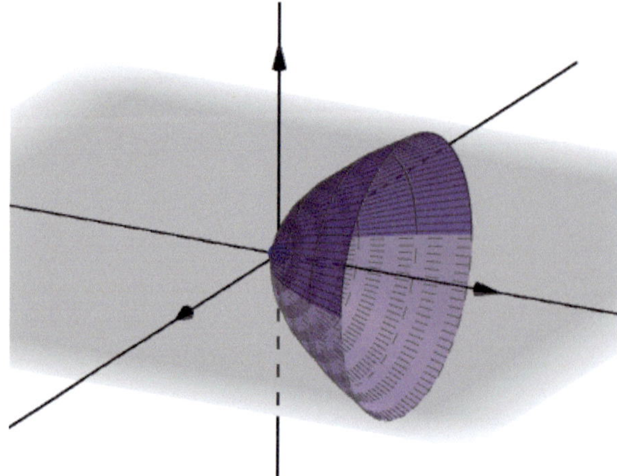

Fig. 6.32 Paraboloid as a solid of rotation

thus

$$1 = \left(1 + \frac{t}{g+1} + \frac{s}{g+1}\right)(g+1)$$

and therefore

$$g = -(t+s).$$

This, however, represents a plane in the coordinate system. This plane, together with the coordinate walls $x = 0$, $y = 0$ and the "base plane" $z = -1$ bounds a pyramid with volume $V = \frac{G \cdot h}{3} = \frac{1}{6}$, because the base area is a triangle with an area ½ and the height is exactly one unit (see Fig. 6.34).

Fig. 6.33 Curved surface

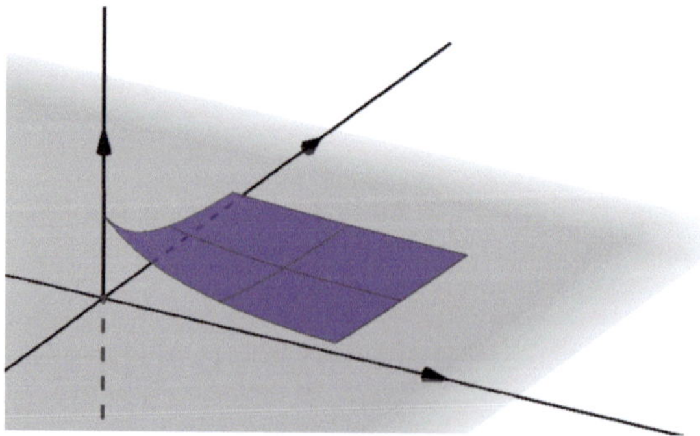

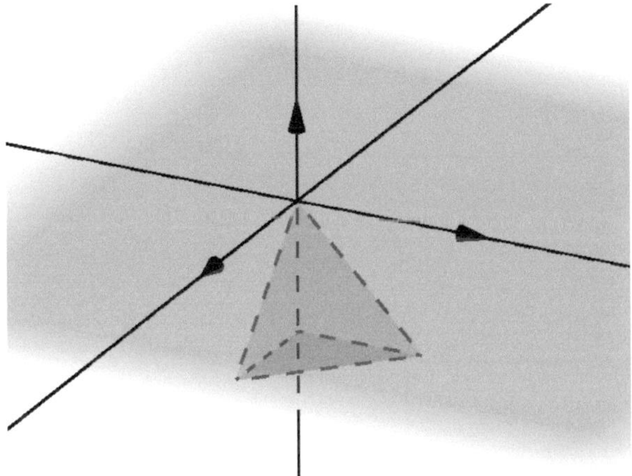

Fig. 6.34 A plane as the fan surface

Thus, the volume under the infinite initial surface is exactly $1/6$ and we have shown

$$\int_{x=0}^{\infty}\int_{y=0}^{\infty} \frac{dydx}{3(1+x+y)^3} = \frac{1}{6}.$$

6.22 The Surface that Belongs to a Pyramid Above the xy-Plane

Again, we turn the tables and now consider the case where the pyramid is the initial body above the xy-plane. The initial surface (see Fig. 6.35) is then

$$f(x, y) = \frac{x+y}{3}.$$

We use the planes $x = 0$, $y = 0$, $z = 0$ and $x + y = 1$ as boundary surfaces for the pyramid.

Then we get

$$f\left(\frac{t}{g+1}, \frac{s}{g+1}\right) = \frac{t+s}{3(g+1)} = \frac{(g+1)^3}{3}$$

for the fan surface. Thus, $(g+1)^4 = t + s$ applies. This surface can be seen in Fig. 6.36 (green). Here, we also need to consider the boundaries $x = 0$ and $y = 0$. The corresponding solid lies below this surface and over the plane $z = -1+x+y$

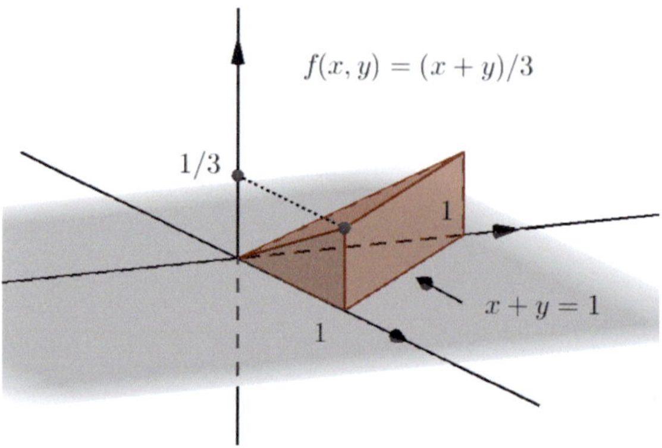

Fig. 6.35 Pyramid above the xy-plane

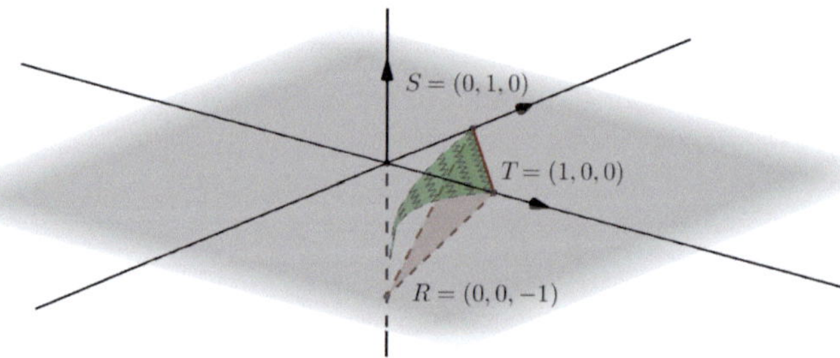

Fig. 6.36 Fan surface of a pyramid

(red) through the points $R = (0, 0, -1)$, $T = (1, 0, 0)$ and $S = (0, 1, 0)$ (see Fig. 6.36). The volume is then calculated to be $1/9$, because the associated pyramid has a rectangular base with area $\sqrt{2} \cdot \frac{1}{3} = \frac{\sqrt{2}}{3}$ and height $\frac{\sqrt{2}}{2}$.

This example reminds us of the fan curve of the linear function in Sect. 6.3.

6.23 The Surface Under the Function $z = 1/\left(1 + x^2 + y^2\right)^3$

Next, we will consider the surface function

$$f(x, y) = \frac{1}{3\left(1 + x^2 + y^2\right)^3}$$

In a somewhat exaggerated representation, the surface is shown in Fig. 3.37.

Here, we have

$$\frac{1}{3\left(1+\left(\frac{t}{g+1}\right)^2+\left(\frac{s}{g+1}\right)^2\right)^3} = f\left(\frac{t}{g+1}, \frac{s}{g+1}\right) = \frac{(g+1)^3}{3},$$

so

$$1 = \left(1+\left(\frac{t}{g+1}\right)^2+\left(\frac{s}{g+1}\right)^2\right)(g+1)$$

and thus

$$g+1 = (g+1)^2+t^2+s^2 \text{ or } \left(g+\frac{1}{2}\right)^2+t^2+s^2 = \frac{1}{4},$$

which for us means a sphere around the center $\left(0,0,-\frac{1}{2}\right)$ with a radius of ½. Therefore, the figure under the initial surface has the same volume as the mentioned sphere, namely $V = \frac{4\pi\left(\frac{1}{2}\right)^3}{3} = \frac{\pi}{6}$ and we have

$$\iint\limits_{xy\text{-plane}} \frac{dxdy}{3(1+x^2+y^2)^3} = \frac{\pi}{6}.$$

In Fig. 6.37, we have slightly beautified the figure to better highlight the shape. This example strongly reminds of the corresponding two-dimensional example by Berendonk in Sect. 6.9.

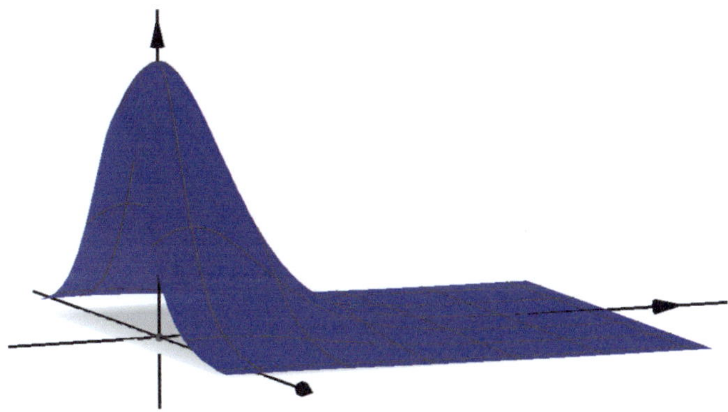

Fig. 6.37 Unbounded surface

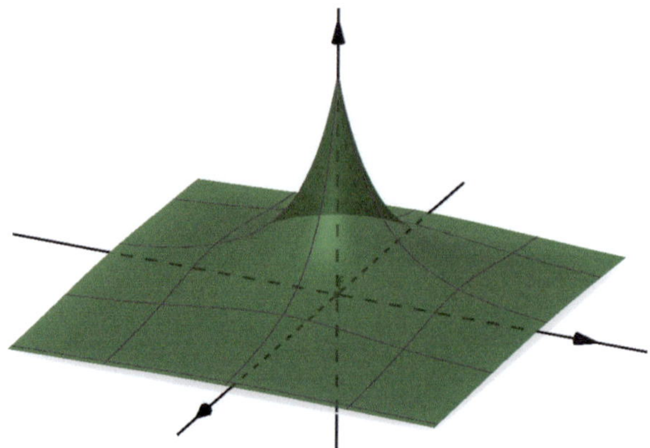

Fig. 6.38 Surface leading to a cone as fan surface

6.24 Parametric Representations in Space

Even here with the area functions, it is worth switching to parametric representations. The parametric representation of a surface always has two free parameters. We choose the variables u and v. The initial area above the xy-plane is then given by

$$x(u, v), y(u, v) \text{ and } z(u, v).$$

The associated fan area then has the representation $t(u, v)$, $s(u, v)$and $g(u, v)$. Our task is now to find a connection between these function triples. The following applies:

$$z(u, v) = \frac{(g(u, v) + 1)^3}{3}, x(u, v) = \frac{t(u, v)}{g(u, v) + 1}, \text{ and } y(u, v) = \frac{s(u, v)}{g(u, v) + 1}.$$

Thus, (x, y, z) can be reconstructed from (t, s, g). On the other hand, we can also calculate (t, s, g) from (x, y, z) because

$$g(u, v) + 1 = \sqrt[3]{3z(u, v)},$$
$$t(u, v) = x(u, v)(g(u, v) + 1) = x(u, v) \cdot \sqrt[3]{3z(u, v)}$$

and

$$s(u, v) = y(u, v)(g(u, v) + 1) = y(u, v) \cdot \sqrt[3]{3z(u, v)}.$$

Thus, we have found a three-dimensional counterpart to Theorem 2.

6.25 The Cone Again

Now we want to apply the above result for parametric surfaces and consider the cone again

$$g = -\sqrt{t^2 + s^2}$$

from Sect. 6.19 below the xy-plane as a fan surface. We use the parameter variables $u = t$ and $v = s$. Then $g + 1 = 1 - \sqrt{t^2 + s^2} = 1 - \sqrt{u^2 + v^2}$. For the initial surface, we then have

$$z = \frac{(g+1)^3}{3} = \frac{\left(1 - \sqrt{u^2 + v^2}\right)^3}{3}, x = \frac{t}{g+1} = \frac{u}{1 - \sqrt{u^2 + v^2}}$$
$$\text{and } y = \frac{s}{g+1} = \frac{v}{1 - \sqrt{u^2 + v^2}}.$$

Thus, we have found a parameterization of the initial surface We have already encountered this surface in Sect. 6.19 (see Fig. 6.28).

Finally, we now show that the formulas for the surface from Sect. 6.19 and those from Sect. 6.25 actually match. We use the parameter equations developed above for z, x and y, namely

$$z = \frac{\left(1 - \sqrt{u^2 + v^2}\right)^3}{3}, x = \frac{u}{1 - \sqrt{u^2 + v^2}}, y = \frac{v}{1 - \sqrt{u^2 + v^2}}.$$

and obtain

$$\begin{aligned} z &= \frac{\left(1 - \sqrt{u^2 + v^2}\right)^3}{3} = \frac{\left(1 - \sqrt{u^2 + v^2}\right)^3}{3\left(1 - \sqrt{u^2 + v^2} + \sqrt{u^2 + v^2}\right)^3} \\ &= \frac{\left(1 - \sqrt{u^2 + v^2}\right)^3}{3\left(1 + \frac{\sqrt{u^2+v^2}}{1-\sqrt{u^2+v^2}}\right)^3\left(1 - \sqrt{u^2 + v^2}\right)^3} \\ &= \frac{1}{3\left(1 + \sqrt{\frac{u^2}{\left(1-\sqrt{u^2+v^2}\right)^2} + \frac{v^2}{\left(1-\sqrt{u^2+v^2}\right)^2}}\right)^3} = \frac{1}{3\left(1 + \sqrt{x^2 + y^2}\right)^3}, \end{aligned}$$

which confirms the calculations from Sect. 6.19.

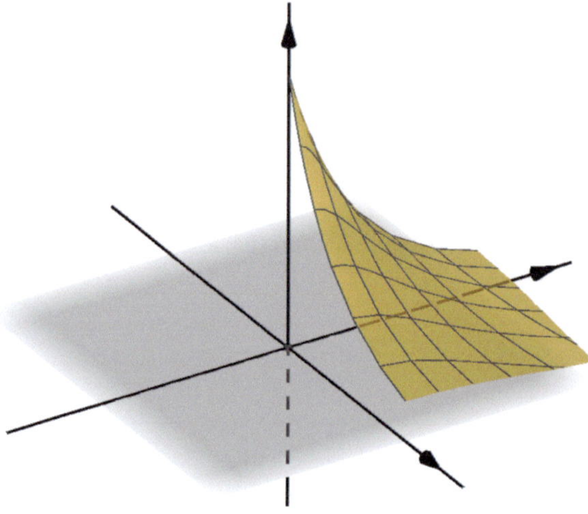

Fig. 6.39 Part of the infinite surface

In Fig. 6.39, we only see a part of the infinite surface. There we can see what the intersection area with the plane $y = 0$ looks like, where we get a sharp peak at the origin because

$$z = \frac{1}{3(1 + |x|)^3}.$$

6.26 The Surface Belonging to a Paraboloid Under the xy-Plane

Our last example is again a paraboloid, as promised above. This time it is located under the xy-plane v. From below it is limited by $z = -1$, see Fig. 6.40.

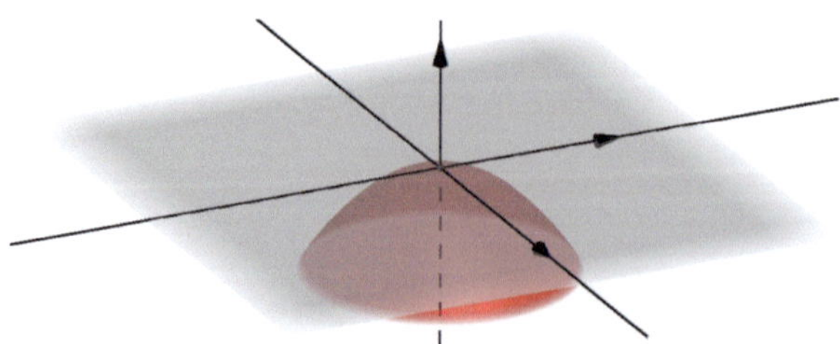

Fig. 6.40 Paraboloid below the xy-plane

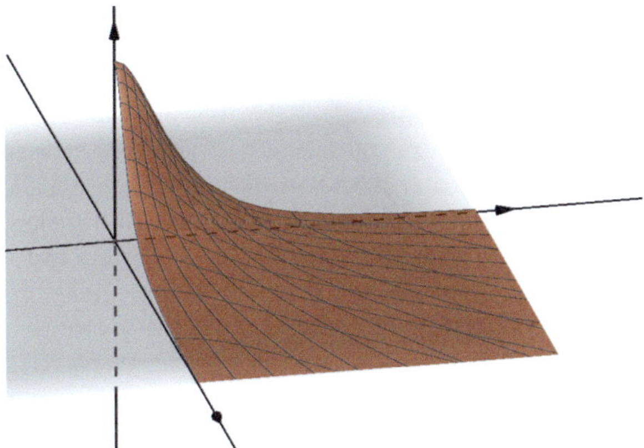

Fig. 6.41 Parametric surface

Here, $g(u, v) = -u^2 - v^2$, so $g + 1 = 1 - u^2 - v^2$. With this, we have the parametric representation of the paraboloid

$$g + 1 = 1 - u^2 - v^2, t = u \quad \text{and} \quad s = v.$$

Now, we can also calculate the parametric representation of the initial surface. We have.

$$z(u, v) = \frac{\left(1 - u^2 - v^2\right)^3}{3}, \; x = \frac{u}{1 - u^2 - v^2}, \quad \text{and} \quad y = \frac{v}{1 - u^2 - v^2}.$$

The corresponding surface can be seen in Fig. 6.41. As a result, we find that the volume under the described surface function is equal to $\frac{\pi}{2}$, just like the volume of the paraboloid below the xy-plane. In Fig. 6.41, we only see the initial surface for $x \geq 0$ and $y \geq 0$, so we can, so to speak, look into the surface. The other quadrants of the plane are of course also relevant for the volume calculation. The steep rise of the surface has been somewhat exaggerated for aesthetic reasons in Fig. 6.41.

6.27 Summary

The fan method gives us the opportunity to handle Riemann sums in many cases and to relate complicated surfaces to simpler and more familiar surfaces. In this way, we can get to know the Riemann integral better and gain deeper insight into its importance. In this process, students can acquire a solid foundation for dealing with calculus.

The exchange of rectangles for triangles is a “translation process”. The ability to translate is often a sign of understanding. If students can practice this process of translation, they may also find the opportunity to expand their understanding and competence for dealing with higher mathematics.

We also noticed that Fermat solved the problem of integrating the Witch of Agnesi purely algebraically using two variable transformations. In this chapter, we have applied such variable transformations several times, but always with a geometric interpretation in mind, where in the two-dimensional case Riemann rectangles were exchanged for triangles, and in the three-dimensional case square columns were exchanged for square pyramids. So, we always had a clear picture that illustrated the transformation to us. Thus, the variable transformation did not remain an algebraic algorithm but became a geometric exchange operation that appealed directly to our understanding and can also be grasped by inexperienced students. In this way, dealing with this chapter can serve as a kind of preparatory work for understanding variable transformations when working with integrals.

Bibliography

Aslaksen H., Kirfel C. (2024), Integration by Riemann triangles, Mathematics Magazine (Mathematical Association of America), Vol. 97, No 1, 2024, pp. 4–22. https://doi.org/10.1080/0025570X.2023.2285389

Fermat P. (1896): Sur la transformation et la simplification des équations de lieux, pour la comparaison sous toutes les formes des aires curvilignes, soit entres elles, soit avec des rectilignes, et en même temps sur l'emploi de la progression géométrique pour la quadrature des paraboles et hyperboles à l'infini. In: Tannery, P., Henry C. Œuvres de Fermat, Tome troisième, Gauthier-Villars et fils, pp. 216–237. English Translation: http://science.larouchepac.com/fermat/

Haftendorn, D. (2017), *Kurven erkunden und verstehen,* Springer Verlag

Kirfel, C. (2020): Die Fächermethode zur Bestimmung von Integralen, Siller, H.-S., Weigel, W. & Wörler, J. F. (Hrsg.). Beiträge zum Mathematikunterricht 2020. Münster: WTM-Verlag, 2020, pp. 497–500. https://doi.org/10.37626/GA9783959871402.0

Lawrence, J. D. (2014). *A catalog of Special Plane Curves.* New York, NY: Dover Publications.

Nelsen, R. B. (2000), Proofs Without Words II, Classroom Materials, Mathematical Association of Amerika.

Paradís J., Pla, J. and Viader P. (2008), *Fermat's Method of Quadrature,* Revue d'histoire des mathématiques 14, pp. 5–51.

Internet Addresses

[Agnesi] https://en.wikipedia.org/wiki/Witch_of_Agnesi

A New Round with π

7

7.1 Introduction

The calculation of the circumference of a circle was already a popular problem in antiquity, and there were different solutions [Approximations to pi] in various cultures.

Already in the second century BCE, Archimedes and Zu Chongzhi [Zu Chongzhi] in the fifth century AD developed good approximations for π, which was very helpful in calculating the circumference U of a circle due to the formula $U = 2\pi r$. Both Archimedes and Zu Chongzhi used geometric methods that involved inscribing (or circumscribing) regular polygons into a circle and thus approximating the circumference of the circle by polygonal circumferences. In Fig. 7.1 we see a regular hexagon that approximates the circle quite well. If the associated circle has a radius of 1, then the hexagon has a circumference of 6 and we see that the value of π must be close to $\frac{6}{2} = 3$. The dodecagon next to it approximates the circle much better. Archimedes managed to describe how the circumference changes when moving to a new polygon with twice as many corners and an angle $\beta = \frac{\alpha}{2}$. He halved the angle in the hexagon several times and finally arrived at a 96-gon and found that $\frac{223}{71} < \pi < \frac{22}{7}$. Written in decimal numbers this becomes $3.1408\ldots < \pi < 3.1428\ldots$, which is a surprisingly good approximation to π.

The Archimedean method for determining π then proceeds as follows: If we start with a regular n-gon inscribed in a circle with a radius of 1, the circumference of the n-gon is $U_n = nT$, where T is the side length of the n-gon as seen in Fig. 7.1 on the left. Now we halve the angle and get a new polygon, this time with $2n$ corners and thus a new approximation for the circumference, namely $U_{2n} = 2nt$, where $t = GB$ is the side length of the $2n$-gon (see Figs. 7.1 and 7.2). Then in Fig. 7.2 it follows that.

© The Author(s), under exclusive license to Springer-Verlag GmbH, DE, part of Springer Nature 2026

C. Kirfel, *Side Paths in the History of Mathematics*, Mathematics Study Resources 21,
https://doi.org/10.1007/978-3-662-72918-2_7

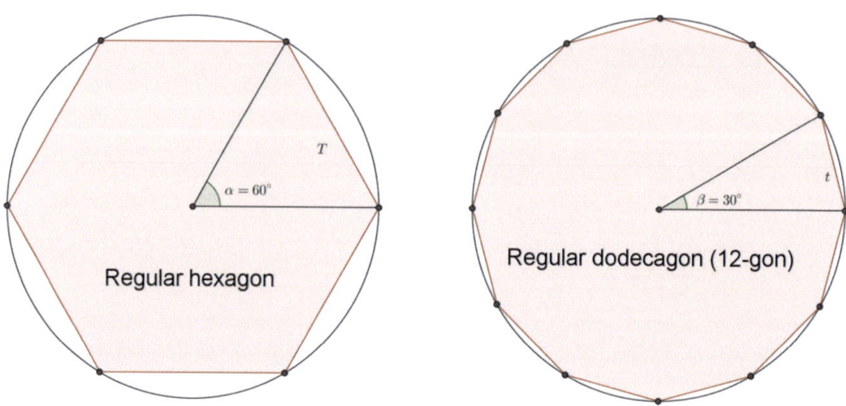

Fig. 7.1 Regular polygons

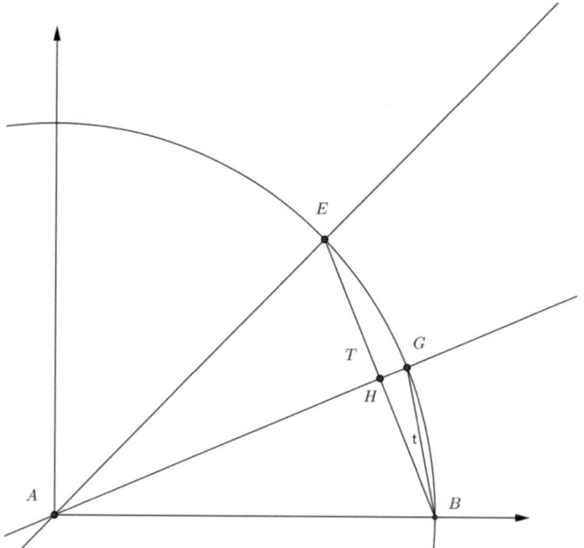

Fig. 7.2 Dividing the angle into two halves

$AH^2 + \left(\frac{T}{2}\right)^2 = 1$ and $(1 - AH)^2 + \left(\frac{T}{2}\right)^2 = t^2$, so $t^2 = 2 - 2\sqrt{1 - \left(\frac{T}{2}\right)^2} = 2 - \sqrt{4 - T^2}$ or

$$t = \sqrt{2 - \sqrt{4 - \mathrm{T}^2}}$$

If you halve the angle several times, as was the case with Archimedes, roots of roots and more and more nested root expressions appear. In this chapter, however, we want to approximate π using rational numbers, i.e., using fractions, and nested root expressions are not a good starting point. Therefore, we leave the Archimedean approach and turn to a completely different method that was developed in the seventeenth century.

First, we need to make some preparations so that we are familiar with the central concepts that will be used in the further course of the chapter. On the one hand, we will often use the tangent function and its inverse function, the arctangent function. Secondly, we need to familiarize ourselves with series, especially the series for the arctangent function. With these two ingredients, we will then be able to achieve interesting results about the number π.

7.2 The Tangent of an Angle

In trigonometry, the tangent of an angle α is defined as the ratio of $\sin(\alpha)$ to $\cos(\alpha)$, i.e., as the ratio of the opposite to the adjacent catheti in a right-angled triangle with angle α, as can be seen in Fig. 7.3.

The right-angled triangles *EBA* and *CDA* are similar and therefore the ratio of the catheti is

$$\tan(\alpha) = \frac{\tan(\alpha)}{1} = \frac{EB}{AB} = \frac{CD}{AD} = \frac{\sin(\alpha)}{\cos(\alpha)}.$$

Here, we can directly measure the tangent of the angle α as the length of the line *EB*. We immediately see that $\tan(0) = 0$, because the sine of $0°$ is zero. Additionally, $\tan(45°) = 1$, because the sine and cosine at $45°$ are equal. The tangent of an angle thus indicates the slope that the angle describes. In the mountains, you may sometimes come across traffic signs indicating the slope of the road, e.g.,

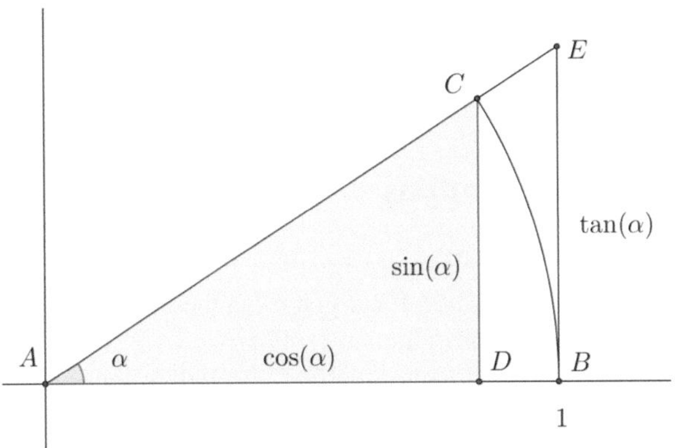

Fig. 7.3 The definition of the tangent of an angle

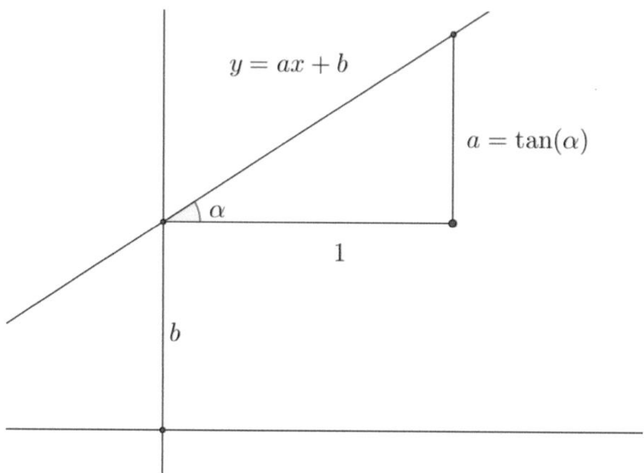

Fig. 7.4 The tangent is the slope of the line

4%. On such a stretch, we have $\tan(\alpha) = 0.04$, i.e., for every 100 m of imagined straight-line distance, it goes down 4 m. Slopes of 8–10% are common. Anything above that is extremely steep. We also encounter the tangent of the angle when specifying formulas for lines in a coordinate system. The tangent of the angle that the line $y = ax + b$ forms with the x-axis or a parallel to the x-axis is precisely the slope a of the line (see Fig. 7.4).

In the following Table 7.1, we see some values of the tangent function for various angles between 0° and 85°.

We easily see that the tangent function increases as the angle (below 90°) increases, and gradually grows towards infinity. If we choose an angle from the first column, we can directly read off the tangent value from the second column, while the third column provides the angle in radians. To specify an angle in radians, we indicate the length of the arc corresponding to that angle along a unit circle (radius 1). For example, an angle of 360° has the radian measure 2π, an angle of 180° has the radian measure π, an angle of 45° has the radian measure $\pi/4$ etc. In Table 7.1, we see, for example, that an angle of 45° has a radian measure of $0.785398 \approx \pi/4$.

7.3 Arctangent Identities

The arctangent function is the inverse function of the tangent function. It helps us to find the corresponding angle for a given tangent value. We just need to use Table 7.1 in reverse. For instance, $\arctan(2.747475) = 1.22173$, corresponds to an angle of 70°. The value 1.22173 is therefore the radian measure of an angle of 70°, i.e., the length of the arc corresponding to an angle of 70° on the unit circle.

Table 7.1 Tangent values

Angle (Degrees)	Tangent	Angle (Radians)
0	0	0
5	0.087489	0.087266
10	0.176327	0.174533
15	0.267949	0.261799
20	0.36397	0.349066
25	0.466308	0.436332
30	0.57735	0.523599
35	0.700207	0.610865
40	0.839099	0.698132
45	1	0.785398
50	1.191753	0.872664
55	1.428147	0.959931
60	1.73205	1.047197
65	2.144506	1.134464
70	2.747475	1.22173
75	3.732047	1.308997
80	5.671272	1.396263
85	11.43001	1.48353

The connection from the arctangent to π is now obvious, because we have

$$\frac{\pi}{4} = \arctan(1),$$

which means that the tangent of 45° is 1, which we have already noted several times. We also have

$$\frac{\pi}{4} = \arctan\left(\frac{1}{2}\right) + \arctan\left(\frac{1}{3}\right).$$

These types of equations are called arctangent identities, and such identities give us the opportunity to construct approximation formulas for π. It is possible to calculate good approximations for the arctangent using the so-called Gregory series. We have already encountered this series in Chap. 6 on the method of Aage Bondesen and also in Chap. 5 on the Resection Method of Leibniz. In Chap. 6 on Aage Bondesen's method, we showed that

$$\arctan(x) = \int_0^x \frac{dt}{1+t^2}.$$

Now we can consider the integrand as a geometric series, because

$$\frac{1}{1+t^2} = 1 - t^2 + t^4 - t^6 + \cdots$$

With this, we can actually perform the integration and we have

$$\arctan(x) = \int_0^x \frac{dt}{1+t^2} = \int_0^x \left(1 - t^2 + t^4 - t^6 + \cdots\right) dt = x - \frac{x^3}{3} + \frac{x^5}{5} - \frac{x^7}{7} \cdots .$$

This, of course, only holds for $|x| < 1$, otherwise the series does not necessarily converge. With this, we have actually obtained a method that allows us to calculate or at least approximate arctangent values. We simply truncate the series after a finite number of terms, for example, after four terms, and thus we have found a value very close to the desired arctangent value, at least when $|x|$ is small. We will later on discuss the size of the errors that may occur in this process.

Of course, the most famous of all arctangent identities is the so-called Gregory-Leibniz formula.

$$\pi = 4\arctan(1) = 4\left(1 - \frac{1}{3} + \frac{1}{5} - \frac{1}{7} + \cdots\right)$$

We have already encountered this identity in Chap. 5 on Leibniz's Resection Method. The beauty of this simple formula is, however, overshadowed by a very slow approximation speed. It takes about 5000 terms to get four correct decimal places after the decimal point for π [5000 terms].

Therefore, it is a good idea to study arctangent identities where the arctangent arguments are less than 1, such as in Eq. (7.1). The smaller they are, the better, because then the approximation speed increases.

Each new arctangent identity gives us the opportunity to calculate π by using the Gregory series for the arctangent function. Therefore, it is exciting to construct new arctangent identities. We can then compare the series and perhaps compete with each other because the series have different rates of convergence and therefore approximate π at different speeds.

In Roger Nelsen's book *Proofs Without Words III* (Nelsen 2016, pp. 75–76), we find the following arctangent identities:

$$\frac{\pi}{4} = \arctan\left(\frac{1}{2}\right) + \arctan\left(\frac{1}{3}\right) \tag{7.1}$$

$$\frac{\pi}{4} = \arctan(3) - \arctan\left(\frac{1}{2}\right) \tag{7.2}$$

$$\frac{\pi}{4} = \arctan(2) - \arctan\left(\frac{1}{3}\right) \tag{7.3}$$

$$\frac{\pi}{2} = \arctan(1) + \arctan\left(\frac{1}{2}\right) + \arctan\left(\frac{1}{3}\right) \tag{7.4}$$

$$\pi = \arctan(1) + \arctan(2) + \arctan(3) \tag{7.5}$$

We now consider the first of the arctangent identities $\frac{\pi}{4} = \arctan\left(\frac{1}{2}\right) + \arctan\left(\frac{1}{3}\right)$. If we use the Gregory series for the arctangent

$$\arctan(x) = x - \frac{x^3}{3} + \frac{x^5}{5} - \frac{x^7}{7} + \cdots,$$

we have found a method for calculating π, since

$$\begin{aligned}
\pi &= 4\left(\arctan\left(\frac{1}{2}\right) + \arctan\left(\frac{1}{3}\right)\right) \\
&= 4\left(\frac{1}{2} - \frac{1}{3}\left(\frac{1}{2}\right)^3 + \frac{1}{5}\left(\frac{1}{2}\right)^5 - \frac{1}{7}\left(\frac{1}{2}\right)^7 + \cdots\right) \\
&\quad + 4\left(\frac{1}{3} - \frac{1}{3}\left(\frac{1}{3}\right)^3 + \frac{1}{5}\left(\frac{1}{3}\right)^5 - \frac{1}{7}\left(\frac{1}{3}\right)^7 + \cdots\right)
\end{aligned}$$

The two series in the brackets converge quickly and with few terms, we already get quite good approximations of π, if we truncate the series after a few terms and neglect the rest. As mentioned above, we will later on conduct detailed investigations of the errors that may occur. The arctangent identities (7.2), (7.3), (7.4), and (7.5) however are unsuitable for the calculation of π since the arguments in the Gregory series are ≥ 1.

7.4 Arctangent Identities with Two Terms

The above identity (7.1) can be easily generalized by studying Fig. 7.5.

In Fig. 7.5, we start with the right-angled triangle ABC, where the horizontal leg has length a while the vertical leg has length 1. Then we have

$$\tan(\alpha) = \frac{1}{a}.$$

Now comes the magic trick. From the "endpoint" $C = (a, 1)$ of the construction, we move diagonally upwards until we hit the line $y = x$ at the point E. For this, we have to move M steps to the left, but also M steps upwards, and then we hit the line $y = x$ at the point E. Therefore, we have

$$a - M = 1 + M$$

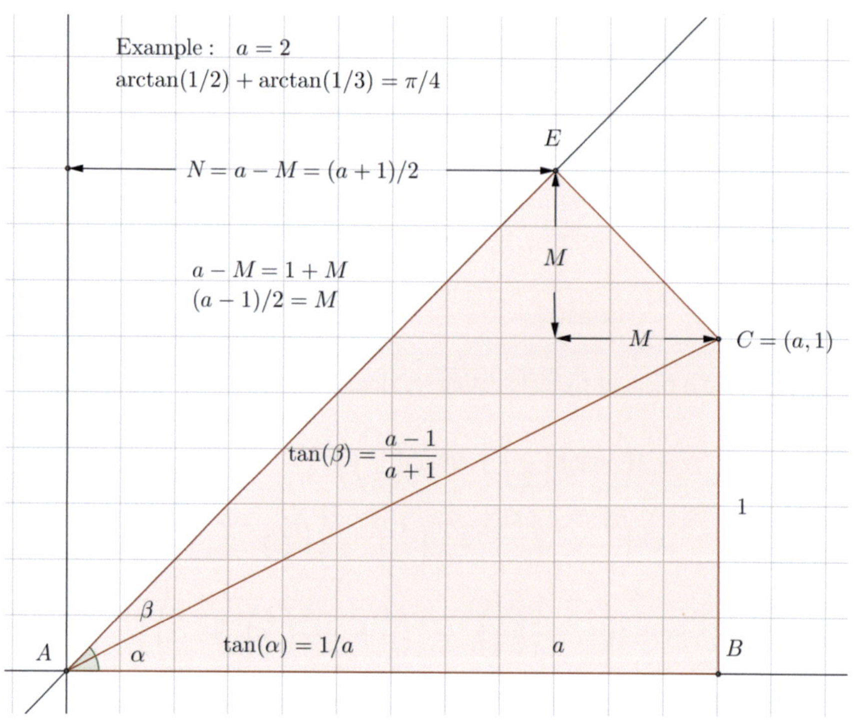

Fig. 7.5 Arctangent identity

$$M = \frac{a-1}{2}$$

and the x-coordinate of E is

$$N = a - M = \frac{a+1}{2}.$$

But now, since AE and EC are perpendicular to each other, we have

$$\tan(\beta) = \frac{EC}{AE} = \frac{M\sqrt{2}}{N\sqrt{2}} = \frac{a-1}{a+1}.$$

All together we get

$$\arctan\left(\frac{1}{a}\right) + \arctan\left(\frac{a-1}{a+1}\right) = \frac{\pi}{4},$$

because the sum of the angles $\alpha + \beta = \pi/4$. In Fig. 7.5, we have chosen $a = 2$.

In this way, we can produce as many new arctangent identities as we want.

The second identity (7.2) from Nelson's initial list

$$\frac{\pi}{4} = \arctan(3) - \arctan\left(\frac{1}{2}\right)$$

can be explained with our formula if we set $a = \frac{1}{3}$. Indeed, we have $\frac{a-1}{a+1} = \frac{-2/3}{4/3} = -\frac{1}{2}$.

We get the third identity (7.3), namely $\frac{\pi}{4} = \arctan(2) - \arctan\left(\frac{1}{3}\right)$, if we set $a = \frac{1}{2}$. The fourth follows from the first and the fact that $\arctan(1) = \frac{\pi}{4}$, and the fifth results when we add the second, third, and fourth.

Here we see that by counting squares, we can create new arctangent identities.

7.5 Three-Term Arctangent Identities

In 1776, Hutton published the following identity [Hutton]:

$$\frac{\pi}{4} = 2 \cdot \arctan\left(\frac{1}{3}\right) + \arctan\left(\frac{1}{7}\right).$$

This was used by Georg von Vega in 1789 to determine π to 143 decimal places, of which the first 126 places were correct.

L.K. Schulz von Strassnitzky (1844) found a similar arctangent identity [Strassnitzky]:

$$\frac{\pi}{4} = \arctan\left(\frac{1}{2}\right) + \arctan\left(\frac{1}{5}\right) + \arctan\left(\frac{1}{8}\right). \tag{7.6}$$

Zacharias Dahse used this formula in 1844 to accurately determine π to 200 decimal places (Nelsen 2016, p. 116).

Remark 7.1

At this point, we have the opportunity to highlight the difference between the Archimedean method and the method with the arctangent identities. While Archimedes used regular polygons to approximate the circle as can be seen in Fig. 7.1 the method based on arctangent identities used non-regular polygons, as can be seen in Fig. 7.6. This figure belongs to L. K. Schulz von Strassnitzky's arctangent identity (7.6). In this figure, $2\pi = 8\alpha + 8\beta + 8\gamma = 8 \cdot \arctan\left(\frac{1}{2}\right) + 8 \cdot \arctan\left(\frac{1}{5}\right) + 8 \cdot \arctan\left(\frac{1}{8}\right)$. The angles that belong to the different side lengths here have rational tangent values. The clear advantage is that calculations can be made rationally.

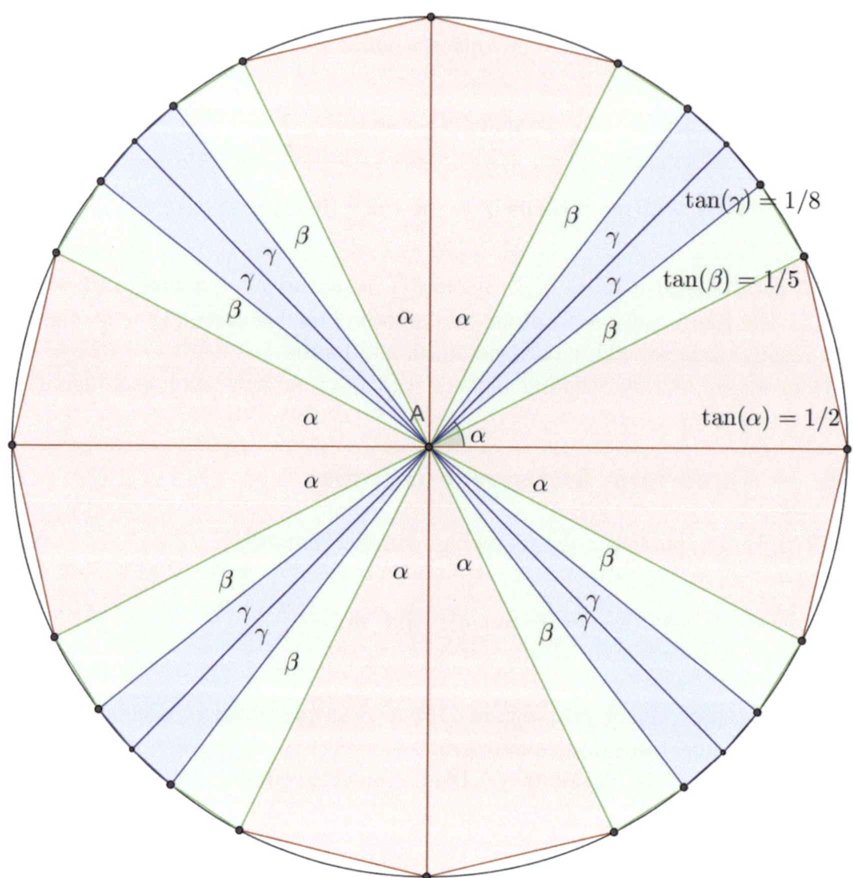

Fig. 7.6 Non-regular polygons

Now we want to generalize the two formulas, those of Hutton and L.K. Schulz von Strassnitzky, which gives us the opportunity to generate many more new arctangent identities. In this way, we can start a race against ourselves.

In Fig. 7.7 we start with the right-angled triangle ABC, where the horizontal leg has the length a, while the vertical one has the length 1. We enlarge this triangle by a factor of b to obtain the triangle $AB'C'$. For the common angle, the following applies

$$\tan(\alpha) = \frac{1}{a}.$$

Now we place a copy of the triangle ABC on top of the construction $C' = (ab, b)$. In the new position, the triangle is then called $C'FD$. Here, the vertical leg has the length a, while the horizontal one has the length 1. Then the hypotenuses in these

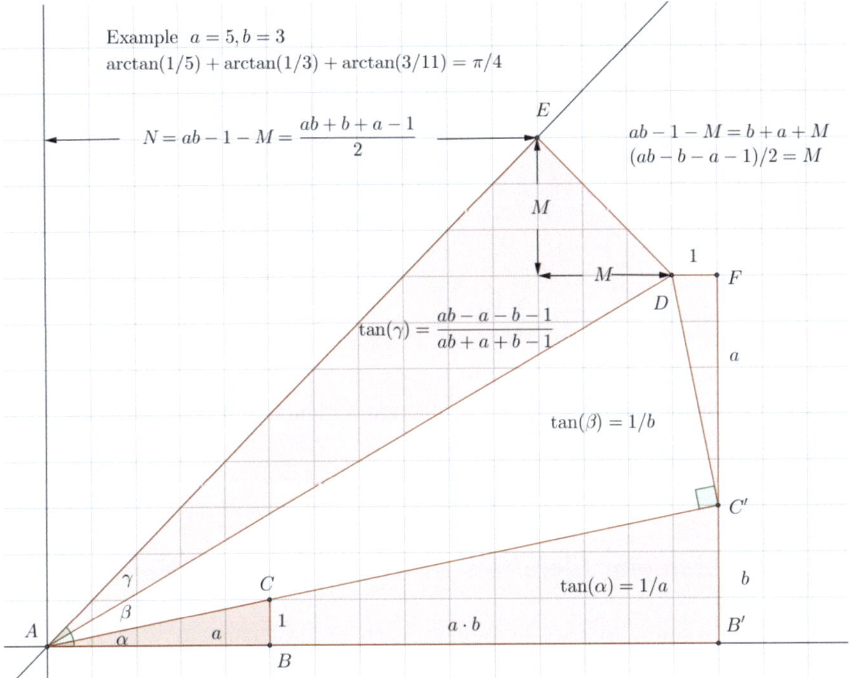

Fig. 7.7 Arctangent identity with three terms

two similar triangles are perpendicular to each other and the ratio of their lengths is equal to the reduction factor

$$\tan(\beta) = \frac{C'D}{AC'} = \frac{1}{b}.$$

Now we use the magic trick again. From the "peak" of the construction $D = (x_D, y_D) = (ab - 1, a + b)$, we move diagonally upwards until we hit the line $y = x$. at point E. In doing so, we have to go M steps to the left and at the same time M steps upwards. We then arrive at point E on the line $y = x$. Therefore, we have

$$\begin{aligned} ab - 1 - M &= a + b + M \text{, i.e.} \\ M &= \frac{ab - a - b - 1}{2}, \end{aligned}$$

and the x-coordinate of E is

$$N = ab - 1 - M = ab - 1 - \frac{ab - a - b - 1}{2} = \frac{ab + a + b - 1}{2}.$$

Generally, we have $x_D - M = y_D + M$ and $N = x_D - M$, which gives us $M = (x_D - y_D)/2$ and $N = (x_D + y_D)/2$. For the tangent of the remaining angle or the "correction angle", this gives us

$$\tan(\gamma) = \frac{M\sqrt{2}}{N\sqrt{2}} = \frac{M}{N} = \frac{x_D - y_D}{x_D + y_D} = \frac{ab - a - b - 1}{ab + a + b - 1}.$$

All together we have

$$\arctan\left(\frac{1}{a}\right) + \arctan\left(\frac{1}{b}\right) + \arctan\left(\frac{ab - a - b - 1}{ab + a + b - 1}\right) = \frac{\pi}{4}.$$

In this way, we can produce as many arctangent identities with three terms as we want. Hutton's formula arises for $a = 3,\ b = 7$ and Strassnitzky's formula for $a = 2,\ b = 5$.

7.6 Arctangent Identities with Three Unit Fractions

Here it can be interesting to investigate when the last tangent value M/N is a unit fraction, so that all three tangent values are unit fractions. We do not know whether M and N are integers, but we do know that $A = 2M$ and $B = 2N$ are integers. The question is therefore when $A = 2M = ab - a - b - 1$ is a divisor of $B = 2N = ab + a + b - 1$. This does not happen too often because B is only slightly larger than A. Now $M/N = A/B = 1$ is not possible and between ½ and 1 there are no unit fractions. Therefore, in our search for unit fractions A/B we can assume that $A/B \leq 1/2$. This is the case when

$$2(ab - a - b - 1) \leq ab + a + b - 1,$$

i.e.,

$$ab - 3(a + b) \leq 1 \quad \text{or} \quad (a - 3)(b - 3) \leq 10.$$

Here only a few values for a and b are possible. Of course, we can exclude $a = 1$ or $b = 1$. We would like to consider the cases where $a = 2$ or $b = 2$, or $a = 3$ or $b = 3$ separately at first.

For $a = 2$ we have $A = 2M = ab - a - b - 1 = b - 3$ and $B = 2N = ab + a + b - 1 = 3b + 1 = 3(b - 3) + 10 = 3A + 10$. If now A/B is to be a unit fraction, then the numerator A must divide the denominator B. Then also A must be a divisor of 10, so $A = 1, 2, 5 \text{ or } 10$. We now go through these cases one by one:

$A = 1$. Here $B = 3A + 10 = 13$. Therefore $\frac{A}{B} = \frac{1}{13}$, which is indeed a unit fraction.

$A = 2$. Here $B = 3A + 10 = 16$. Therefore $\frac{A}{B} = \frac{2}{16} = \frac{1}{8}$.
$A = 5$. Here $B = 3A + 10 = 25$. Therefore $\frac{A}{B} = \frac{5}{25} = \frac{1}{5}$.
$A = 10$. Here $B = 3A + 10 = 40$. Therefore $\frac{A}{B} = \frac{10}{40} = \frac{1}{4}$.

In all cases, we get unit fractions. This gives us the two arctangent identities

$$\frac{\pi}{4} = \arctan\left(\frac{1}{2}\right) + \arctan\left(\frac{1}{4}\right) + \arctan\left(\frac{1}{13}\right),$$

corresponding to $A = 1$ or $A = 10$ and

$$\frac{\pi}{4} = \arctan\left(\frac{1}{2}\right) + \arctan\left(\frac{1}{5}\right) + \arctan\left(\frac{1}{8}\right),$$

Corresponding to $A = 2$ or $A = 5$. The former is new, but the latter formula is precisely that of L. K. Schulz von Strassnitzky.

For $a = 3$ the following holds $A = 2M = ab - a - b - 1 = 2(b - 2)$ and $B = 2N = ab + a + b - 1 = 4(b - 2) + 10 = 2A + 10$. If now again A/B is to be a unit fraction, then A must again be a divisor of 10, so again $A = 1, 2, 5$ or 10. Because of $A = 2(b - 2)$, A must be even and thus the cases $A = 1$ and $A = 5$ are eliminated. We now go through the remaining two cases one by one:

$A = 2$. Here $B = 2N = 2A + 10 = 14$, so $\frac{A}{B} = \frac{2}{14} = \frac{1}{7}$, is a unit fraction.
$A = 10$. Here $B = 2N = 2A + 10 = 30$, so $\frac{A}{B} = \frac{10}{30} = \frac{1}{3}$ is also a unit fraction.

This gives us in both cases the same arctangent identity

$$\frac{\pi}{4} = \arctan\left(\frac{1}{3}\right) + \arctan\left(\frac{1}{7}\right) + \arctan\left(\frac{1}{3}\right),$$

which we have already seen as the arctangent identity of Hutton.

Finally, we now study the values $a \geq 4$ and $b \geq 4$ in Table 7.2. We also know that

$(a - 3)(b - 3) \leq 10$, so $a \geq 14$ or $b \geq 14$ is not possible.

In Table 7.2, unit fractions appear at four places, but they do not yield any new identities. We have shown that the three mentioned identities are the only ones with three terms where all arguments are unit fractions.

7.7 Arctangent Identities with Multiple Terms

Now we want to generalize the above ideas to also handle more than two or three angles. We start again with the same construction as in Fig. 7.7, until we reach the point $D = (x_D, y_D)$. Now we enlarge the entire construction by the factor c. This preserves all tangent values. On the vertex $D' = (c \cdot x_D, c \cdot y_D)$ of the

Table 7.2 The values of $M/N = (ab - a - b - 1)/(ab + a + b - 1)$ with varying a and b

$b \backslash a$	4	5	6	7	8	9	10	11	12	13	14
4	$\frac{7}{23}$	$\frac{10}{28}$	$\frac{13}{33}$	$\frac{16}{38}$	$\frac{19}{43}$	$\frac{22}{48}$	$\frac{25}{53}$	$\frac{28}{58}$	$\frac{31}{63}$	$\mathbf{\frac{34}{68} = \frac{1}{2}}$	$\frac{37}{73} > \frac{1}{2}$
5	$\frac{10}{28}$	$\frac{14}{34}$	$\frac{18}{40}$	$\frac{22}{46}$	$\mathbf{\frac{26}{52} = \frac{1}{2}}$	$\frac{30}{58} > \frac{1}{2}$					
6	$\frac{13}{33}$	$\frac{18}{40}$	$\frac{23}{47}$	$\frac{28}{54} > \frac{1}{2}$							
7	$\frac{16}{38}$	$\frac{22}{46}$	$\frac{28}{54} > \frac{1}{2}$								
8	$\frac{19}{43}$	$\mathbf{\frac{26}{52} = \frac{1}{2}}$									
9	$\frac{22}{48}$	$\frac{30}{58} > \frac{1}{2}$									
10	$\frac{25}{53}$										
11	$\frac{28}{58}$										
12	$\frac{31}{63}$										
13	$\mathbf{\frac{34}{68} = \frac{1}{2}}$										
14	$\frac{37}{73} > \frac{1}{2}$										

construction, we now place a copy of the triangle. ADZ, where $Z = (x_D, 0)$ is the foot of the perpendicular from D to the x-axis. In this triangle, the lengths of the legs are x_D and y_D. This ensures that for the angle γ at the origin in the new triangle $AD'D''$ we have $\tan(\gamma) = \frac{1}{c}$. The new "peak" of the construction is now $D'' = (cx_D - y_D, cy_D + x_D)$. From there, we move diagonally towards the point W on the line $y = x$ (Fig. 7.8). If $\alpha + \beta + \gamma > 45°$, then $M < 0$ and therefore also $\arctan(M/N) < 0$.

In this way, we find

$$abc - c - a - b - M = ab + bc + ac - 1 + M$$

$$M = \frac{abc - ab - ac - bc - a - b - c + 1}{2}$$

and the x-coordinate of D'' is

$$N = abc - c - a - b - M = \frac{abc + ab + ac + bc - a - b - c - 1}{2}.$$

For the remaining angle δ, we have

$$\tan(\delta) = \frac{M\sqrt{2}}{N\sqrt{2}} = \frac{abc - ab - ac - bc - a - b - c + 1}{abc + ab + ac + bc - a - b - c - 1}.$$

All in all, this gives

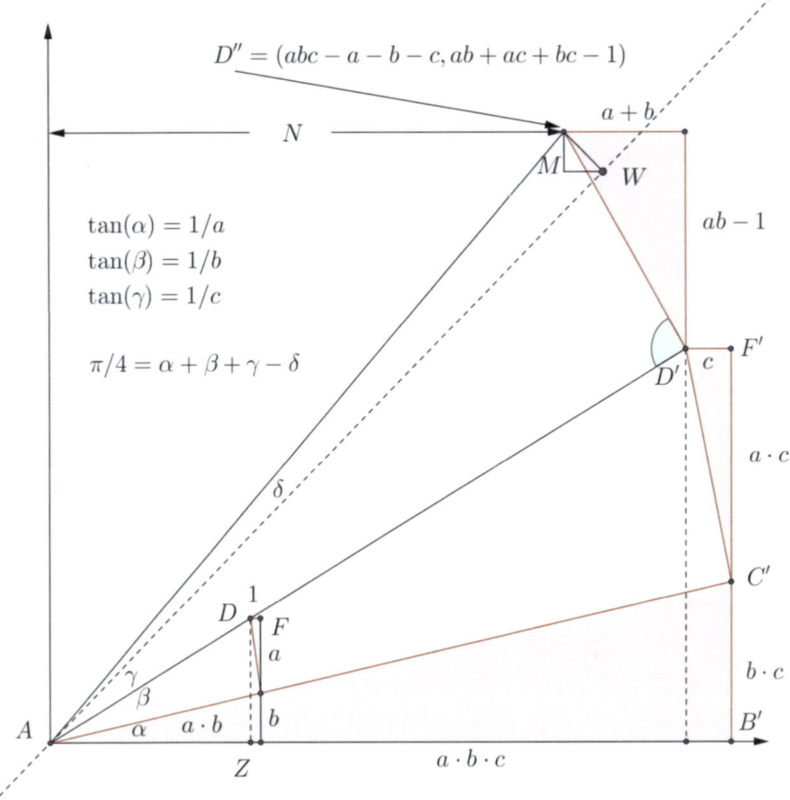

Fig. 7.8 Four-term arctangent identity

$$\arctan\left(\frac{1}{a}\right) + \arctan\left(\frac{1}{b}\right) + \arctan\left(\frac{1}{c}\right)$$
$$+ \arctan\left(\frac{abc - ab - ac - bc - a - b - c + 1}{abc + ab + ac + bc - a - b - c - 1}\right) = \frac{\pi}{4}.$$

In this way, we can produce as many four-term arctangent identities as we want, where we can freely choose three of the angles (or their tangent values).

This procedure describes an algorithm that we can apply when we want to add any number of angles, the tangent values of which are known.

- Start with the first angle, where the tangent value is known, and construct the corresponding right-angled triangle.
- Call the "peak" of the construction $D = (x_D, y_D)$ and the corresponding foot on the x-axis $Z = (x_D, 0)$.

- Enlarge the entire construction with the reciprocal of the next tangent value (e.g. with b)
- On top of the peak $D' = (bx_D, by_D)$ of the construction, we now place a copy of the triangle ADZ that is rotated 90°. Here A is the origin and Z is the foot on the x-axis perpendicular below D. The new peak is then $P = (bx_D - y_D, by_D + x_D)$.
- Complete the construction by moving diagonally from P towards the line $y = x$ and calculate the tangent of the remaining angle:

$$\frac{M}{N} = \frac{bx_D - y_D - by_D - x_D}{bx_D - y_D + by_D + x_D} = \frac{(b-1)x_D - (b+1)y_D}{(b+1)x_D + (b-1)y_D}$$

- Write down the new arctangent identity.

When adding four free angles, we get

$$x_D = abcd - ab - ac - ad - bc - bd - cd + 1$$

and

$$y_D = abc + abd + acd + bcd - a - b - c - d,$$

which gives us

$$\frac{M}{N} = \frac{abcd - abc - abd - acd - bcd - ab - ac - ad - bc - bd - cd + a + b + c + d + 1}{abcd + abc + abd + acd + bcd - ab - ac - ad - bc - bd - cd - a - b - c - d + 1}$$

and thus we have

$$\arctan\left(\frac{1}{a}\right) + \arctan\left(\frac{1}{b}\right) + \arctan\left(\frac{1}{c}\right) + \arctan\left(\frac{1}{d}\right) + \arctan\left(\frac{M}{N}\right) = \frac{\pi}{4}.$$

Similarly, for five free angles, we get

$$\begin{aligned} x_D = {} & abcde - abc - abd - abe - acd - ace - ade \\ & - bcd - bce - bde - cde - a - b - c - d - e \end{aligned}$$

and

$$\begin{aligned} y_D = {} & abcd + abce + acde + bcde + abde - ab - ac \\ & - ad - bc - bd - cd - ae - be - ce - de + 1. \end{aligned}$$

If we choose $a = b = c = d$ in the case of four free angles, we get

$$D = (x_D, y_D) = \left(a^4 - 6a^2 + 1, 4a^3 - 4a\right)$$

and if we choose $a = b = c = d = e$ for five free angles, we get

$$D = (x_D, y_D) = \left(a^5 - 10a^2 + 5a, 5a^4 - 10a^2 + 1\right),$$

7.8 Addition of Angles

The addition of angles is strongly related to the multiplication of complex numbers. When two complex numbers are multiplied together, the angles of the individual numbers add up to the angle of the product. If we multiply $a + i$ with $b + i$, we obtain

$$(a + i)(b + i) = ab - 1 + i(a + b)$$

and we see that the product as a complex number corresponds to the point $D = (ab - 1, a + b)$ in Fig. 7.7. This means that the process just described represents nothing more than the multiplication of the complex numbers $a+i, b+i, c+i, \ldots$ This also means that the coordinates of the point $D = (x_D, y_D)$ in our construction can easily be calculated by complex multiplication. Multiples of angles can also be treated in this way.

Because of

$$(a + i)^n = a^n - \binom{n}{2}a^{n-2} + \binom{n}{4}a^{n-4} - \cdots$$
$$+ i\left(\binom{n}{1}a^{n-1} - \binom{n}{3}a^{n-3} + \binom{n}{5}a^{n-5} - \cdots\right)$$

we have

$$x_D = a^n - \binom{n}{2}a^{n-2} + \binom{n}{4}a^{n-4} - \cdots$$

and

$$y_D = \binom{n}{1}a^{n-1} - \binom{n}{3}a^{n-3} + \binom{n}{5}a^{n-5} - \cdots.$$

Now we fill up the angle $n \cdot \arctan(\frac{1}{a})$ until we reach the line $y = x$ by moving diagonally from $D = (x_D, y_D)$ to the line $y = x$. For the remaining angle, we then get

$$M = \frac{x_D - y_D}{2} \text{ and } N = \frac{x_D + y_D}{2},$$

and our identity then reads as follows

$$\frac{\pi}{4} = n \cdot \arctan\left(\frac{1}{a}\right) + \arctan\left(\frac{M}{N}\right).$$

Again, we can produce as many new arctangent identities as we want.

7.9 Machin's Formula

For $n = 4$ and $a = 5$ things turn out particularly well. We have

$$x_D = 5^4 - \binom{4}{2}5^2 + 1 = 625 - 150 + 1 = 476, \text{ and } y_D = 4 \cdot 5^3 - 4 \cdot 5 = 480,$$

so we get the following

$$\frac{M}{N} = \frac{x_D - y_D}{x_D + y_D} = \frac{476 - 480}{476 + 480} = \frac{-1}{239},$$

which compared to $1/a = 1/5$ is incredibly small. For this reason, we can expect that the associated arctangent identity

$$\frac{\pi}{4} = 4 \cdot \arctan\left(\frac{1}{5}\right) + \arctan\left(\frac{-1}{239}\right)$$

will give us an exceptionally good approximation formula for π [Machin 1]. We will see that this is indeed the case. This identity was already discovered in 1706 by the English mathematician John Machin.

Wikipedia [Machin 2] writes:

> "Machin's formula remained the primary tool of π-hunters for centuries (well into the computer era)."

We now use the Gregory series for the arctangent

$$\arctan(x) = x - \frac{x^3}{3} + \frac{x^5}{5} - \frac{x^7}{7} + \cdots$$

For $x = 1/5$ the error R_1, after we have considered four terms of the series, is

$$|R_1| = \left|\frac{x^9}{9} - \frac{x^{11}}{11} + \cdots\right| \le \left|\frac{x^9}{9}\right| = \frac{1}{5^9 \cdot 9} = \frac{1}{17{,}578{,}125} < 5.7 \cdot 10^{-8},$$

because we can arrange the terms that come after $\frac{x^9}{9}$ in pairs, each of which contributes negatively, since we have

$$-\frac{x^{11}}{11} + \frac{x^{13}}{13} < 0,$$

because $11x^2 < 13$, since $|x| < 1$. In general, $-\frac{x^{2k+1}}{2k+1} + \frac{x^{2k+3}}{2k+3} < 0$ because $(2k+1)x^2 < 2k+3$, which in turn follows from $|x| < 1$. The sum of the pairs is then of course also negative and we have shown that

$$R_1 = \frac{x^9}{9} - \frac{x^{11}}{11} + \cdots \le \frac{x^9}{9}.$$

On the other hand, we easily see that

$$R_1 = \frac{x^9}{9} - \frac{x^{11}}{11} + \cdots \geq 0,$$

which again follows from the argument with the pairs. All together we get

$$|R_1| = \left| \frac{x^9}{9} - \frac{x^{11}}{11} \pm \cdots \right| \leq \left| \frac{x^9}{9} \right|.$$

For $x = -1/239$ we only use a single term of the series. The error R_2 can then be estimated as follows:

$$|R_2| = \left| \frac{x^3}{3} - \frac{x^5}{5} + \frac{x^7}{7} \pm \cdots \right| \leq \left| \frac{x^3}{3} \right| = \frac{1}{3 \cdot 239^3} < 2.442 \cdot 10^{-8}.$$

Because of

$$\pi = 16 \arctan\left(\frac{1}{5}\right) - 4 \arctan\left(\frac{1}{239}\right)$$

we can conclude that

$$\left| \pi - \left(\frac{16}{5} - \frac{16}{3 \cdot 5^3} + \frac{16}{5 \cdot 5^5} - \frac{16}{7 \cdot 5^7} - \frac{4}{239} \right) \right| \leq 16|R_1| + 4|R_2|$$
$$= 16 \cdot 5{,}689 \cdot 10^{-8} + 4 \cdot 2{,}442 \cdot 10^{-8} = 100.792 \cdot 10^{-8} < 1.008 \cdot 10^{-6}$$

and the difference between our approximation

$$\frac{16}{5} - \frac{16}{3 \cdot 5^3} + \frac{16}{5 \cdot 5^5} - \frac{16}{7 \cdot 5^7} - \frac{4}{239} = \frac{1{,}231{,}847{,}548}{392{,}109{,}375} = 3.141591674516836\ldots$$

and π is less than $1.008 \cdot 10^{-6}$, which means we have determined π to at least 5 correct decimal places. The sixth decimal place differs by at most one unit from the correct one. The value of π accurate to 20 decimal places is actually

$$\pi = 3.14159265358979323846\ldots.$$

However, we were very lucky here. Because M/N was so small, we only needed to use a single term of the arctangent series. What would happen if we were to increase a from 5 to 10? It turns out that $8 \cdot \arctan(1/10)$ just adds up to a little more than 45°.

Therefore, we have to perform the following calculation for $a = 10$:

$$x_D = a^8 - \binom{8}{6} a^6 + \binom{8}{4} a^4 - \binom{8}{2} a^2 + 1$$

$$= 10^8 - 28 \cdot 10^6 + 70 \cdot 10^4 - 28 \cdot 10^2 + 1 = 72{,}697{,}201$$

$$y_D = 8a^7 - \binom{8}{5}a^5 + \binom{8}{3}a^3 - \binom{8}{1}a$$
$$= 8 \cdot 10^7 - 56 \cdot 10^5 + 56 \cdot 10^3 - 8 \cdot 10 = 74{,}455{,}920.$$

With $M = (x_D - y_D)/2$ and $N = (x_D + y_D)/2$ we then get

$$\frac{M}{N} = -\frac{1{,}758{,}719}{147{,}153{,}121} \approx -0.012,$$

and we obtain

$$\pi = 32 \arctan\left(\frac{1}{10}\right) - 4 \arctan\left(\frac{1{,}758{,}719}{147{,}153{,}121}\right).$$

Here the approximation

$$32\left(\frac{1}{10} - \frac{1}{3 \cdot 10^3} + \frac{1}{5 \cdot 10^5} - \frac{1}{7 \cdot 10^7}\right) - \frac{4 \cdot 1{,}758{,}719}{147{,}153{,}121} = 3.14159037$$

is even worse than for $a = 5$ and $n = 4$. This is due to the following: Although the error R_1 for $\arctan\left(\frac{1}{10}\right)$ in this case is $< \frac{x^9}{9} = \frac{1}{10^9 \cdot 9} < 1.2 \cdot 10^{-10}$, the error for the correction term $\arctan\left(\frac{1{,}758{,}719}{147{,}153{,}121}\right)$ is much larger, namely

$$R_2 = \frac{x^3}{3} - \frac{x^5}{5} + \frac{x^7}{7} - \cdots > \frac{x^3}{3} - \frac{x^5}{5} > 5 \cdot 10^{-7}.$$

The total error is greater than $4R_2 > 2 \cdot 10^{-6}$ and the approximation is worse than in the case of $n = 4, a = 5$, where the error was below $1.008 \cdot 10^{-6}$.

Of course, we could have used more than a single term of the series for the correction term. However, this would lead to calculations with third powers of $N = 147{,}153{,}121$ and thus to calculations with huge numbers.

The reason why we do not get a significantly better approximation is that the correction term $M/N \approx -0.012$ is much larger than in the case where $a = 5$ and this counts much more than the reduction of the original angle, where the value of the tangent of $1/5$ decreases to $1/10$. This also means that increasing a does not necessarily lead to a better approximation formula. It is important to keep an eye on the tangent value M/N of the correction angle and ensure that small values occur here.

7.10 Advances with the Help of Computers

We have shown here how to produce identities of the form

$$\frac{\pi}{4} = n_1 \cdot \arctan\left(\frac{1}{a_1}\right) + n_2 \cdot \arctan\left(\frac{1}{a_2}\right) + \cdots + n_k \cdot \arctan\left(\frac{1}{a_k}\right) + \arctan\left(\frac{M}{N}\right)$$

by multiplying complex numbers

$$(a_1 + i)^{n_1}(a_2 + i)^{n_2} \cdots (a_k + i)^{n_k}$$

and in the end using a correction term, as we have done many times. In our case, the last tangent value is not necessarily a unit fraction. Our method was to calculate the coordinates of the vertex of our construction $D = (x_D, y_D)$. Then we have $M = (x_D - y_D)/2$ and $N = (x_D + y_D)/2$ and we know the tangent value M/N of the correction angle.

In the course of the previous section we realized that we need to look for small values of M/N. The value M/N is a measure of the quality of the arctangent identity, as it is crucial for the order of magnitude of the error in the term $\arctan(M/N)$. The smaller M/N is, the better the approximation for π will be. We choose the values $1/a_i$ ourselves, so the resulting errors are under our control.

By systematically testing various values of a_i and n_i and then calculating the corresponding correction angles, or their tangent values M/N, we can potentially discover new effective arctangent identities, much like what happened with Machin's identity. This is best accomplished with a computer program. Once we find a small value for M/N (compared to the reciprocals of the $a_1, a_2, \cdots a_k$), we are in luck and can expect a good approximation formula for π, where relatively little work is needed (with few terms from the Gregory series).

We first consider the formula

$$\frac{\pi}{4} = n \cdot \arctan\left(\frac{1}{a}\right) + \arctan\left(\frac{M}{N}\right) \tag{7.7}$$

and go through the relevant values of a and n. Machin's identity for $a = 5$ and $n = 4$ is a special case of this formula. With our program, we first search the area $2 \le a \le 200$. We start with a value for a and then successively calculate the multiples of the angle, $n \cdot \arctan(1/a)$, until the total angle exceeds the 45° mark (tangent = slope 1), see Fig. 7.9. We are interested in the last value before the total angle exceeds 45° and the first one after. This means that in Fig. 7.9, both angles SOV and VOT are of interest. If one of them is small, it may potentially lead to good arctangent identities.

Now we calculate the tangent of the correction angle M/N. If this value is smaller than the previous tangent values of the correction angles, we ask the program to print both the tangent value of the correction angle and the associated values of a, n, x_D, and y_D. Then we move on to the next value for a. In this way, the program always prints the smallest tangent value of the correction angle so far.

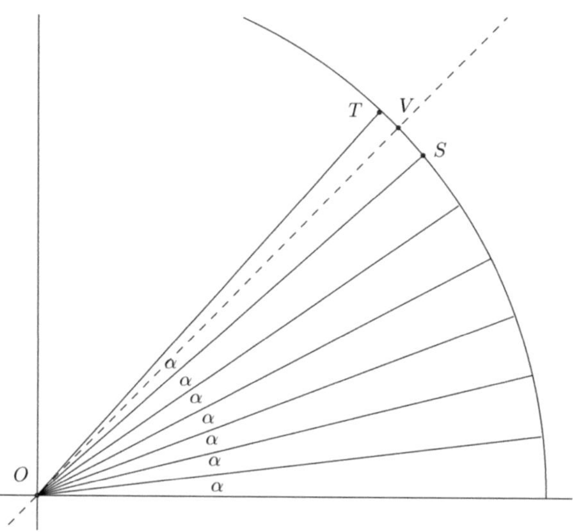

Fig. 7.9 Multiples of an angle

Computerprogram (Python)

```
Import math;
        import numpy as np
        a = 2
        Min = 1
        while a < 201:
            x = a;y = 1;n = 1
            while x > y:
                r = a*x–y;s = x + a*y
                x = r; y = s;
                n = n + 1
                M = (x–y)/2
                N = (x + y)/2
                if abs(M/N) < abs(Min):
                  Min = M/N
                  print("x = ", x, " y = ”", y, "n = ",n, "a = ", a)
                  print("Minimum = ",Min)
                  N3 = a*a*a
                  N5 = N3*a*a
                  N7 = N5*a*a
                  P = 4*n*(1/a-1/(3*N3) + 1/(5*N5)-1/(7*N7)) + 4*M/N)
                  print("pi = ",P)
                a = a + 1
```

7.11 Explanations of the Program

We go through the values from 2 to 200 for the variable a. The ends of the loop can be easily changed. After that, we ask how many times an angle with $\tan(\alpha) = 1/a$ must be added until the sum approaches the 45° mark. To do this, we simply form the powers $(a + i)^n$ and ask when the result first lies above the line $y = x$. For this, we calculate

$$r = a * x - y;\ s = x + a * y$$

$$x = r;\ y = s;$$

We now calculate the tangent of the correction angle $M/N = (x_D - y_D)/(x_D + y_D)$.

If this tangent value is smaller than all previous corresponding tangent values (if abs(M/N) < abs(Min)), the program prints out the associated values. In a later version of the program, we instead use the bound $1.76 \cdot 10^{-5}$ at this point (if abs(M/N) < 1.76*10^−5). Both the tangent value of the last correction angle before the 45° mark and the first after the 45° mark are small angles and one of the two may possibly provide a new minimum value, which is then displayed if necessary, see Fig. 7.9. The associated values for a and n, as well as the values for x_D and y_D are also printed.

The results produced by the program are summarized in the following Table 7.3:

After the initial starting value for $a = 2$, which corresponds to Eq. (7.1), and $a = 4$ we get Machin's formula with $a = 5$ and $n = 4$ in line 4. Afterwards, the values $a = 14$ and $n = 11$ are optimal and finally, the combination $a = 28$ and $n = 22$ surprises with an incredibly small correction angle, whose tangent is just $M/N = 1.77 \cdot 10^{-5}$. We will investigate this situation in detail. After that, no better correction angles occur up to and including $a = 218$. We also see that the values for x_D and y_D increase rapidly.

Table 7.3 Results of the computer program

a	n	x_D y_D	M/N
2	1	$x_D = 2$ $y_D = 1$	0.33
2	2	$x_D = 3$ $y_D = 4$	− 0.14
4	3	$x_D = 52$ $y_D = 47$	0.05
5	4	$x_D = 476$ $y_D = 480$	− 0.0014
14	11	$x_D = 2{,}947{,}746{,}030{,}782$ $y_D = 2{,}941{,}761{,}989{,}627$	− 0.001
28	22	$x_D = 49{,}324{,}070{,}120{,}846{,}121{,}302{,}260{,}022{,}403{,}183$ $y_D = 49{,}322{,}325{,}613{,}363{,}940{,}973{,}893{,}167{,}838{,}056$	$1.77 \cdot 10^{-5}$

a = 28, n = 22
xM = 49,324,070,120,846,121,302,260,022,403,183
yM = 49,322,325,613,363,940,973,893,167,838,056
M/N = 1.7684452322826654e − 05

These results astonish us because the correction angle is so small. Among all angles α with $\tan(\alpha) = 1/a$ for $2 \le a \le 218$ the smallest correction angle appears already at $a = 28$. In a certain sense this gives us the best approximation of π. This is an astonishing result. It also turns out that

$$22 \cdot \arctan\left(\frac{1}{28}\right) = 44.9989867^\circ.$$

This means the following: When we calculate $(28 + i)^{22}$, the point $D = (x_D, y_D)$ will come very, very close to the line $y = x$. But we also know how we can calculate the tangent value of the correction angle, namely

$$\tan(\beta) = \frac{x_D - y_D}{x_D + y_D} < 1.77 \cdot 10^{-5}.$$

Since the correction angle here is so tiny, the approximation for π is particularly good.

We get the following approximation for π:

$$22 \cdot 4\left(\frac{1}{28} - \frac{1}{3 \cdot 28^3} + \frac{1}{5 \cdot 28^5} - \frac{1}{7 \cdot 28^7}\right) - \frac{4 \cdot (x_D - y_D)}{x_D + y_D} = 3.141592653588877,$$

where the error is less than 10^{-12} and we have 11 correct digits after the decimal point, as you can see, when comparing the result with the first 20 correct digits of π. As already mentioned above, we have

$$\pi = 3.14159265358979323846\ldots.$$

Here we have again used four terms from the Gregory series to calculate $\arctan(\frac{1}{28})$. This means that we get the following error term R_1:

$$|R_1| = \left|\frac{x^9}{9} - \frac{x^{11}}{11} + \cdots\right| < \left|\frac{x^9}{9}\right| = \frac{1}{28^9 \cdot 9} < 1.06 \cdot 10^{-14}.$$

For the second arctangent value in (7.7), we again used only a single term from the series. This results in

$$|R_2| = \left|\frac{x^3}{3} - \frac{x^5}{5} + \cdots\right| \leq \left|\frac{x^3}{3}\right| < \frac{1.77^3 \cdot 10^{-15}}{3} < 1.85 \cdot 10^{-15}.$$

All in all, the total error is then at most

$$|4 \cdot 22 \cdot R_1 + 4 \cdot R_2| < 1.06 \cdot 88 \cdot 10^{-14} + 4 \cdot 1.85 \cdot 10^{-15} < 10^{-12},$$

which confirms the above calculations. The approximation value calculated by Archimedes $22/7$ for π naturally gives the approximation value $22/28$ for $\pi/4$, which places our discovery in the historical context.

7.12 Expansion of the Search Range

Next, we expand the range for a to $219 \leq a \leq 27{,}000$. Understandably, new good values appear, which can be read off in Table 7.4. The associated values for x_D and y_D are not listed, as they are numbers with many hundreds, sometimes even thousands of digits. In the range $1{,}809 \leq a \leq 27{,}000$ we could not find any further improvements.

The last line in Table 7.4 gives us $a = 1{,}808$, $n = 1{,}420$ and $M/N = 1.33 \cdot 10^{-8}$. If we again use four terms from the Gregory series to calculate $\arctan(1/1{,}808)$, this results in an error

$$|R_1| = \left|\frac{x^9}{9} - \frac{x^{11}}{11} \pm \cdots\right| < \left|\frac{x^9}{9}\right| = \frac{1}{1{,}808^9 \cdot 9} < 3.04 \cdot 10^{-27}.$$

For the second arctangent value, we again use only a single term from the series. This results in

$$|R_2| = \left|\frac{x^3}{3} - \frac{x^5}{5} \pm \cdots\right| < \left|\frac{x^3}{3}\right| < \frac{1.33^3 \cdot 10^{-24}}{3} < 7.84 \cdot 10^{-23}.$$

Table 7.4 Results of the computer program when expanding the search range

a	n	M/N
219	172	$1.54 \cdot 10^{-5}$
233	183	$-4.73 \cdot 10^{-6}$
452	355	$1.21 \cdot 10^{-6}$
904	710	$2.53 \cdot 10^{-7}$
1356	1065	$7.56 \cdot 10^{-8}$
1,808	1,420	$1.33 \cdot 10^{-8}$

All in all, the total error is then at most

$$|4 \cdot 1{,}420 \cdot R_1 + 4 \cdot R_2| < 3.04 \cdot 5{,}680 \cdot 10^{-27} + 4 \cdot 7.84 \cdot 10^{-23} < 10^{-21},$$

and we could determine π to 20 digits of accuracy in this way. Along the way, however, we had to calculate the values for x_D and y_D, which contain many thousands of digits. To be able to calculate the values M and N, we must calculate $(1{,}808 + i)^{1{,}420}$, a number whose real part and imaginary part contain roughly 5,000 digits. Given these enormous calculations, the gain—about 10 new digits of π—is rather modest compared to the case $a = 28, n = 22$ and $M/N = 1.77 \cdot 10^{-5}$.

Remark 7.2
In Table 7.4, the sequence of values $a = 452$, $a = 904 = 2 \cdot 452$, $a = 1{,}356 = 3 \cdot 452$ and $a = 1{,}808 = 4 \cdot 452$ is also surprising. This sequence of multiples of 452 presents us with a real puzzle. For the two initial values that seem to fall out of the pattern, $a = 219$ and $a = 233$, we surprisingly get $219 + 233 = 452$. In Table 7.6, we see that the tangent value for $a = 2{,}260 = 5 \cdot 452$ is also very small, but not smaller than for $a = 1{,}808$. Therefore, it is not registered as a new minimum by the program. The values of n follow a similar pattern. From line 3 onwards, they always increase by 355 units per new line.

If we extend the search for new minima to the area $27{,}000 \leq a < 36{,}000$, new minima appear, which we have collected in Table 7.5.

Here, the values x_D and y_D have around 100,000 digits. In this range, we are entirely dependent on calculations with the computer. Interestingly, a similar pattern appears here as in Table 7.4. The a values increase by 452 units from line to line, while the values for n increase by 355 units from line to line.

Table 7.5 Results of the extended search

a	n	M/N
27,791	21,827	$1.32 \cdot 10^{-8}$
28,243	22,182	$1.19 \cdot 10^{-8}$
28,695	22,537	$1.07 \cdot 10^{-8}$
29,147	22,892	$9.52 \cdot 10^{-9}$
29,599	23,247	$8.35 \cdot 10^{-9}$
30,051	23,602	$7.22 \cdot 10^{-9}$
30,503	23,957	$6.12 \cdot 10^{-9}$
30,955	24,312	$5.05 \cdot 10^{-9}$
31,407	24,667	$4.01 \cdot 10^{-9}$
31,859	25,022	$3.01 \cdot 10^{-9}$
32,311	25,377	$2.03 \cdot 10^{-9}$
32,763	25,732	$1.07 \cdot 10^{-9}$
33,215	26,087	$1.54 \cdot 10^{-10}$

Table 7.6 Results of the computer program after modifying the condition for the output

$a = k_1 \cdot 219 + k_2 \cdot 14$	$n = k_1 \cdot 172 + k_2 \cdot 11$	M/N
$28 = 0 \cdot 219 + 2 \cdot 14$	$22 = 0 \cdot 172 + 2 \cdot 11$	$1.76 \cdot 10^{-5}$
$219 = 1 \cdot 219 + 0 \cdot 14$	$172 = 1 \cdot 172 + 0 \cdot 11$	$1.54 \cdot 10^{-5}$
$233 = 1 \cdot 219 + 1 \cdot 14$	$183 = 1 \cdot 172 + 1 \cdot 11$	$-4.73 \cdot 10^{-6}$
$438 = 2 \cdot 219 + 0 \cdot 14$	$344 = 2 \cdot 172 + 0 \cdot 11$	$1.14 \cdot 10^{-5}$
$452 = 2 \cdot 219 + 1 \cdot 14$	$355 = 2 \cdot 172 + 2 \cdot 11$	$1.21 \cdot 10^{-6}$
$466 = 2 \cdot 219 + 2 \cdot 14$	$366 = 2 \cdot 172 + 2 \cdot 11$	$-8.35 \cdot 10^{-6}$
$480 = 2 \cdot 219 + 3 \cdot 14$	$377 = 2 \cdot 172 + 3 \cdot 11$	$-1.73 \cdot 10^{-5}$
$657 = 3 \cdot 219 + 0 \cdot 14$	$516 = 3 \cdot 172 + 0 \cdot 11$	$1.06 \cdot 10^{-5}$
$671 = 3 \cdot 219 + 1 \cdot 14$	$527 = 3 \cdot 172 + 1 \cdot 11$	$3.81 \cdot 10^{-6}$
$685 = 3 \cdot 219 + 2 \cdot 14$	$538 = 3 \cdot 172 + 2 \cdot 11$	$-2.73 \cdot 10^{-6}$
$699 = 3 \cdot 219 + 3 \cdot 14$	$549 = 3 \cdot 172 + 3 \cdot 11$	$-9.02 \cdot 10^{-6}$
$713 = 3 \cdot 219 + 4 \cdot 14$	$560 = 3 \cdot 172 + 4 \cdot 11$	$-1.50 \cdot 10^{-5}$
$862 = 4 \cdot 219 - 1 \cdot 14$	$677 = 4 \cdot 172 - 1 \cdot 11$	$1.56 \cdot 10^{-5}$
$876 = 4 \cdot 219 + 0 \cdot 14$	$688 = 4 \cdot 172 + 0 \cdot 11$	$1.03 \cdot 10^{-5}$
$890 = 4 \cdot 219 + 1 \cdot 14$	$699 = 4 \cdot 172 + 1 \cdot 11$	$5.23 \cdot 10^{-6}$
$904 = 4 \cdot 219 + 2 \cdot 14$	$710 = 4 \cdot 172 + 2 \cdot 11$	$2.53 \cdot 10^{-7}$
$918 = 4 \cdot 219 + 3 \cdot 14$	$721 = 4 \cdot 172 + 3 \cdot 11$	$-4.57 \cdot 10^{-6}$
$932 = 4 \cdot 219 + 4 \cdot 14$	$732 = 4 \cdot 172 + 4 \cdot 11$	$-9.26 \cdot 10^{-6}$
$946 = 4 \cdot 219 + 5 \cdot 14$	$743 = 4 \cdot 172 + 5 \cdot 11$	$-1.38 \cdot 10^{-5}$
$1{,}081 = 5 \cdot 219 - 1 \cdot 14$	$849 = 5 \cdot 172 - 1 \cdot 11$	$1.44 \cdot 10^{-5}$
$1{,}095 = 5 \cdot 219 + 0 \cdot 14$	$860 = 5 \cdot 172 + 0 \cdot 11$	$1.02 \cdot 10^{-5}$
$1{,}109 = 5 \cdot 219 + 1 \cdot 14$	$871 = 5 \cdot 172 + 1 \cdot 11$	$6.13 \cdot 10^{-6}$
$1{,}123 = 5 \cdot 219 + 2 \cdot 14$	$882 = 5 \cdot 172 + 2 \cdot 11$	$2.11 \cdot 10^{-6}$
$1{,}137 = 5 \cdot 219 + 3 \cdot 14$	$893 = 5 \cdot 172 + 3 \cdot 11$	$-1.80 \cdot 10^{-6}$
$1{,}151 = 5 \cdot 219 + 4 \cdot 14$	$904 = 5 \cdot 172 + 4 \cdot 11$	$-5.63 \cdot 10^{-6}$
$1{,}165 = 5 \cdot 219 + 5 \cdot 14$	$915 = 5 \cdot 172 + 5 \cdot 11$	$-9.36 \cdot 10^{-6}$
$1{,}179 = 5 \cdot 219 + 6 \cdot 14$	$926 = 5 \cdot 172 + 6 \cdot 11$	$-1.30 \cdot 10^{-5}$
$1{,}193 = 5 \cdot 219 + 7 \cdot 14$	$937 = 5 \cdot 172 + 7 \cdot 11$	$-1.65 \cdot 10^{-5}$
$1{,}286 = 6 \cdot 219 - 2 \cdot 14$	$1{,}010 = 6 \cdot 172 - 2 \cdot 11$	$1.72 \cdot 10^{-5}$
$1{,}300 = 6 \cdot 219 - 1 \cdot 14$	$1{,}021 = 6 \cdot 172 - 1 \cdot 11$	$1.37 \cdot 10^{-5}$
$1{,}314 = 6 \cdot 219 + 0 \cdot 14$	$1{,}032 = 6 \cdot 172 + 0 \cdot 11$	$1.01 \cdot 10^{-5}$
$1{,}328 = 6 \cdot 219 + 1 \cdot 14$	$1{,}043 = 6 \cdot 172 + 1 \cdot 11$	$6.74 \cdot 10^{-6}$
$1{,}342 = 6 \cdot 219 + 2 \cdot 14$	$1{,}054 = 6 \cdot 172 + 2 \cdot 11$	$3.37 \cdot 10^{-6}$
$1{,}356 = 6 \cdot 219 + 3 \cdot 14$	$1{,}065 = 6 \cdot 172 + 3 \cdot 11$	$7.56 \cdot 10^{-8}$
$1{,}370 = 6 \cdot 219 + 4 \cdot 14$	$1{,}076 = 6 \cdot 172 + 4 \cdot 11$	$-3.15 \cdot 10^{-6}$
$1{,}384 = 6 \cdot 219 + 5 \cdot 14$	$1{,}087 = 6 \cdot 172 + 5 \cdot 11$	$-6.32 \cdot 10^{-6}$

(continued)

Table 7.6 (continued)

$a = k_1 \cdot 219 + k_2 \cdot 14$	$n = k_1 \cdot 172 + k_2 \cdot 11$	M/N
$1{,}398 = 6 \cdot 219 + 6 \cdot 14$	$1{,}098 = 6 \cdot 172 + 6 \cdot 11$	$-9.42 \cdot 10^{-6}$
$1{,}412 = 6 \cdot 219 + 7 \cdot 14$	$1{,}109 = 6 \cdot 172 + 7 \cdot 11$	$-1.24 \cdot 10^{-5}$
$1{,}426 = 6 \cdot 219 + 8 \cdot 14$	$1{,}120 = 6 \cdot 172 + 8 \cdot 11$	$-1.54 \cdot 10^{-5}$
$1{,}505 = 7 \cdot 219 - 2 \cdot 14$	$1{,}182 = 7 \cdot 172 - 2 \cdot 11$	$1.62 \cdot 10^{-5}$
$1{,}519 = 7 \cdot 219 - 1 \cdot 14$	$1{,}193 = 7 \cdot 172 - 1 \cdot 11$	$1.31 \cdot 10^{-5}$
$1{,}533 = 7 \cdot 219 + 0 \cdot 14$	$1{,}204 = 7 \cdot 172 + 0 \cdot 11$	$1.01 \cdot 10^{-5}$
$1{,}547 = 7 \cdot 219 + 1 \cdot 14$	$1{,}215 = 7 \cdot 172 + 1 \cdot 11$	$7.19 \cdot 10^{-6}$
$1{,}561 = 7 \cdot 219 + 2 \cdot 14$	$1{,}226 = 7 \cdot 172 + 2 \cdot 11$	$4.29 \cdot 10^{-6}$
$1{,}575 = 7 \cdot 219 + 3 \cdot 14$	$1{,}237 = 7 \cdot 172 + 3 \cdot 11$	$1.44 \cdot 10^{-6}$
$1{,}589 = 7 \cdot 219 + 4 \cdot 14$	$1{,}248 = 7 \cdot 172 + 4 \cdot 11$	$-1.35 \cdot 10^{-6}$
$1{,}603 = 7 \cdot 219 + 5 \cdot 14$	$1{,}259 = 7 \cdot 172 + 5 \cdot 11$	$-4.10 \cdot 10^{-6}$
$1{,}617 = 7 \cdot 219 + 6 \cdot 14$	$1{,}270 = 7 \cdot 172 + 6 \cdot 11$	$-6.80 \cdot 10^{-6}$
$1{,}631 = 7 \cdot 219 + 7 \cdot 14$	$1{,}281 = 7 \cdot 172 + 7 \cdot 11$	$-9.46 \cdot 10^{-6}$
$1{,}645 = 7 \cdot 219 + 8 \cdot 14$	$1{,}292 = 7 \cdot 172 + 8 \cdot 11$	$-1.20 \cdot 10^{-5}$
$1{,}659 = 7 \cdot 219 + 9 \cdot 14$	$1{,}303 = 7 \cdot 172 + 9 \cdot 11$	$-1.46 \cdot 10^{-5}$
$1{,}673 = 7 \cdot 219 + 10 \cdot 14$	$1{,}314 = 7 \cdot 172 + 10 \cdot 11$	$-1.71 \cdot 10^{-5}$
$1{,}724 = 8 \cdot 219 - 2 \cdot 14$	$1{,}354 = 8 \cdot 172 - 2 \cdot 11$	$1.54 \cdot 10^{-5}$
$1{,}738 = 8 \cdot 219 - 1 \cdot 14$	$1{,}365 = 8 \cdot 172 - 1 \cdot 11$	$1.27 \cdot 10^{-5}$
$1{,}752 = 8 \cdot 219 + 0 \cdot 14$	$1{,}376 = 8 \cdot 172 + 0 \cdot 11$	$1.01 \cdot 10^{-5}$
$1{,}766 = 8 \cdot 219 + 1 \cdot 14$	$1{,}387 = 8 \cdot 172 + 1 \cdot 11$	$7.53 \cdot 10^{-6}$
$1{,}780 = 8 \cdot 219 + 2 \cdot 14$	$1{,}398 = 8 \cdot 172 + 2 \cdot 11$	$4.98 \cdot 10^{-6}$
$1{,}794 = 8 \cdot 219 + 3 \cdot 14$	$1{,}409 = 8 \cdot 172 + 3 \cdot 11$	$2.48 \cdot 10^{-6}$
$1{,}808 = 8 \cdot 219 + 4 \cdot 14$	$1{,}420 = 8 \cdot 172 + 4 \cdot 11$	$1.33 \cdot 10^{-8}$
$1{,}822 = 8 \cdot 219 + 5 \cdot 14$	$1{,}431 = 8 \cdot 172 + 5 \cdot 11$	$-2.41 \cdot 10^{-6}$
$1{,}836 = 8 \cdot 219 + 6 \cdot 14$	$1{,}442 = 8 \cdot 172 + 6 \cdot 11$	$-4.80 \cdot 10^{-6}$
$1{,}850 = 8 \cdot 219 + 7 \cdot 14$	$1{,}453 = 8 \cdot 172 + 7 \cdot 11$	$-7.16 \cdot 10^{-6}$
$1{,}864 = 8 \cdot 219 + 8 \cdot 14$	$1{,}464 = 8 \cdot 172 + 8 \cdot 11$	$-9.48 \cdot 10^{-6}$
$1{,}878 = 8 \cdot 219 + 9 \cdot 14$	$1{,}475 = 8 \cdot 172 + 9 \cdot 11$	$-1.17 \cdot 10^{-5}$
$1{,}892 = 8 \cdot 219 + 10 \cdot 14$	$1{,}486 = 8 \cdot 172 + 10 \cdot 11$	$-1.40 \cdot 10^{-5}$
$1{,}906 = 8 \cdot 219 + 11 \cdot 14$	$1{,}497 = 8 \cdot 172 + 11 \cdot 11$	$-1.62 \cdot 10^{-5}$
$1{,}929 = 9 \cdot 219 - 3 \cdot 14$	$1{,}515 = 9 \cdot 172 - 3 \cdot 11$	$1.72 \cdot 10^{-5}$
$1{,}943 = 9 \cdot 219 - 2 \cdot 14$	$1{,}526 = 9 \cdot 172 - 2 \cdot 11$	$1.48 \cdot 10^{-5}$
$1{,}957 = 9 \cdot 219 - 1 \cdot 14$	$1{,}537 = 9 \cdot 172 - 1 \cdot 11$	$1.24 \cdot 10^{-5}$
$1{,}971 = 9 \cdot 219 + 0 \cdot 14$	$1{,}548 = 9 \cdot 172 + 0 \cdot 11$	$1.01 \cdot 10^{-5}$
$1{,}985 = 9 \cdot 219 + 1 \cdot 14$	$1{,}559 = 9 \cdot 172 + 1 \cdot 11$	$7.80 \cdot 10^{-6}$
$1{,}999 = 9 \cdot 219 + 2 \cdot 14$	$1{,}570 = 9 \cdot 172 + 2 \cdot 11$	$5.53 \cdot 10^{-6}$

(continued)

Table 7.6 (continued)

$a = k_1 \cdot 219 + k_2 \cdot 14$	$n = k_1 \cdot 172 + k_2 \cdot 11$	M/N
$2{,}013 = 9 \cdot 219 + 3 \cdot 14$	$1{,}581 = 9 \cdot 172 + 3 \cdot 11$	$3.29 \cdot 10^{-6}$
$2{,}027 = 9 \cdot 219 + 4 \cdot 14$	$1{,}592 = 9 \cdot 172 + 4 \cdot 11$	$1.08 \cdot 10^{-6}$
$2{,}041 = 9 \cdot 219 + 5 \cdot 14$	$1{,}603 = 9 \cdot 172 + 5 \cdot 11$	$-1.08 \cdot 10^{-6}$
$2{,}055 = 9 \cdot 219 + 6 \cdot 14$	$1{,}614 = 9 \cdot 172 + 6 \cdot 11$	$-3.23 \cdot 10^{-6}$
$2{,}069 = 9 \cdot 219 + 7 \cdot 14$	$1{,}625 = 9 \cdot 172 + 7 \cdot 11$	$-5.35 \cdot 10^{-6}$
$2{,}083 = 9 \cdot 219 + 8 \cdot 14$	$1{,}636 = 9 \cdot 172 + 8 \cdot 11$	$-7.44 \cdot 10^{-6}$
$2{,}097 = 9 \cdot 219 + 9 \cdot 14$	$1{,}647 = 9 \cdot 172 + 9 \cdot 11$	$-9.50 \cdot 10^{-6}$
$2{,}111 = 9 \cdot 219 + 10 \cdot 14$	$1{,}658 = 9 \cdot 172 + 10 \cdot 11$	$-1.15 \cdot 10^{-5}$
$2{,}125 = 9 \cdot 219 + 11 \cdot 14$	$1{,}669 = 9 \cdot 172 + 11 \cdot 11$	$-1.35 \cdot 10^{-5}$
$2{,}139 = 9 \cdot 219 + 12 \cdot 14$	$1{,}680 = 9 \cdot 172 + 12 \cdot 11$	$-1.55 \cdot 10^{-5}$
$2{,}153 = 9 \cdot 219 + 13 \cdot 14$	$1{,}691 = 9 \cdot 172 + 13 \cdot 11$	$-1.74 \cdot 10^{-5}$
$2{,}148 = 10 \cdot 219 - 3 \cdot 14$	$1{,}687 = 10 \cdot 172 - 3 \cdot 11$	$1.64 \cdot 10^{-5}$
$2{,}162 = 10 \cdot 219 - 2 \cdot 14$	$1{,}698 = 10 \cdot 172 - 2 \cdot 11$	$1.43 \cdot 10^{-5}$
$2{,}176 = 10 \cdot 219 - 1 \cdot 14$	$1{,}709 = 10 \cdot 172 - 1 \cdot 11$	$1.21 \cdot 10^{-5}$
$2{,}190 = 10 \cdot 219 + 0 \cdot 14$	$1{,}720 = 10 \cdot 172 + 0 \cdot 11$	$1.00 \cdot 10^{-5}$
$2{,}204 = 10 \cdot 219 + 1 \cdot 14$	$1{,}731 = 10 \cdot 172 + 1 \cdot 11$	$8.01 \cdot 10^{-6}$
$2{,}218 = 10 \cdot 219 + 2 \cdot 14$	$1{,}742 = 10 \cdot 172 + 2 \cdot 11$	$5.97 \cdot 10^{-6}$
$2{,}232 = 10 \cdot 219 + 3 \cdot 14$	$1{,}753 = 10 \cdot 172 + 3 \cdot 11$	$3.95 \cdot 10^{-6}$
$2{,}246 = 10 \cdot 219 + 4 \cdot 14$	$1{,}764 = 10 \cdot 172 + 4 \cdot 11$	$1.95 \cdot 10^{-6}$
$2{,}260 = 10 \cdot 219 + 5 \cdot 14$	$1{,}775 = 10 \cdot 172 + 5 \cdot 11$	$-1.54 \cdot 10^{-8}$
$2{,}274 = 10 \cdot 219 + 6 \cdot 14$	$1{,}786 = 10 \cdot 172 + 6 \cdot 11$	$-1.96 \cdot 10^{-6}$
$2{,}288 = 10 \cdot 219 + 7 \cdot 14$	$1{,}797 = 10 \cdot 172 + 7 \cdot 11$	$-3.88 \cdot 10^{-6}$
$2{,}302 = 10 \cdot 219 + 8 \cdot 14$	$1{,}808 = 10 \cdot 172 + 8 \cdot 11$	$-5.78 \cdot 10^{-6}$
$2{,}316 = 10 \cdot 219 + 9 \cdot 14$	$1{,}819 = 10 \cdot 172 + 9 \cdot 11$	$-7.65 \cdot 10^{-6}$
$2{,}330 = 10 \cdot 219 + 10 \cdot 14$	$1{,}830 = 10 \cdot 172 + 10 \cdot 11$	$-9.51 \cdot 10^{-6}$
$2{,}344 = 10 \cdot 219 + 11 \cdot 14$	$1{,}841 = 10 \cdot 172 + 11 \cdot 11$	$-1.13 \cdot 10^{-5}$
$2{,}358 = 10 \cdot 219 + 12 \cdot 14$	$1{,}852 = 10 \cdot 172 + 12 \cdot 11$	$-1.31 \cdot 10^{-5}$
$2{,}372 = 10 \cdot 219 + 13 \cdot 14$	$1{,}863 = 10 \cdot 172 + 13 \cdot 11$	$-1.49 \cdot 10^{-5}$
$2{,}386 = 10 \cdot 219 + 14 \cdot 14$	$1{,}874 = 10 \cdot 172 + 14 \cdot 11$	$-1.67 \cdot 10^{-5}$
$2{,}400 = 10 \cdot 219 + 15 \cdot 14$	$1{,}885 = 10 \cdot 172 + 15 \cdot 11$	$-1.84 \cdot 10^{-5}$*

* The value $a = 2{,}400$ in the last row of this Table is there merely to achieve a "round" conclusion. The absolute value of the corresponding tangent is $-1.84 \cdot 10^{-5}$ and thus not below our desired bound $1.77 \cdot 10^{-5}$

As mentioned in the introduction the Chinese mathematician and astronomer Zu Chongzhi also worked on approximations of π. He found the approximation $355/113$ for π, a result that is astonishing close. This approximation got an own name, Milü. The patterns in our table are based on the numbers 355 and $452 = 4 \cdot 113$ and we can see the strong connection between Zu Chongzhi's result and our findings. The factor 4 stems from the fact that our method aims at calculation $\pi/4$ and not π itself [Milü].

7.13 Second Approach

We now want to make a second attempt and not register new minima in the program, but all tangent values $M/N < 1.77 \cdot 10^{-5}$ with $a \leq 2400$. The value $1.77 \cdot 10^{-5}$ corresponds to the tangent value that occurred for $a = 28, n = 22$. So, we want to investigate all cases where the tangent value is comparable with the situation $a = 28, n = 22$. For this, only one line must be exchanged in the program:

In stead of "if abs(M/N) < abs(Min):" we have to write "if abs(M/N) < 0.0000177:"

The results, i.e. the values for a, n and M/N, where $|M/N| < 1.77 \cdot 10^{-5}$ are collected in Table 7.6.

Remark 7.3
Table 7.6 also puzzles us. We see that the values of a and n follow certain patterns. It seems that

$$a = k_1 \cdot 219 + k_2 \cdot 14 \text{ and } n = k_1 \cdot 172 + k_2 \cdot 11,$$

where k_1 apparently runs through the natural numbers, while we can observe $-3 \leq k_2 \leq 15$. Within a block, i.e., at a fixed value of k_1, the tangent values of the remainder angle decrease towards the center of the block and then increase again.

The value $946 = 4 \cdot 219 + 5 \cdot 14$ appears in the table. The subsequent value, however, $960 = 4 \cdot 219 + 6 \cdot 14$ does not. But even for 960, we get a good tangent value, namely $1.82 \cdot 10^{-5}$. However, this value is not below our desired bound $1.77 \cdot 10^{-5}$. By changing this bound in the program, we can expand or restrict the occurring patterns.

In the preceding discussions, we have seen that among all arctangent identities of the form

$$\frac{\pi}{4} = n \cdot \arctan\left(\frac{1}{a}\right) + \arctan\left(\frac{M}{N}\right),$$

the one with $a = 28$ and $n = 22$ is quite special and gives us an extraordinary approximation to π. This is a very special result. Additionally, we have found mysterious patterns in the "good" identities that we have not yet been able to explain.

7.14 Attempt to Explain the Emerging Pattern

Above, in Remark 7.3, we saw that the multiples of $a = 219$ are particularly good starting points for approximations of π. We want to show here that at least a doubling will always lead to fairly good approximations. For this, we study the following helpful arctangent identity:

$$\arctan\left(\frac{1}{a}\right) = 2\arctan\left(\frac{1}{2a}\right) - \arctan\left(\frac{1}{4a^3 + 3a}\right). \tag{7.8}$$

We immediately see that this identity can be used to improve others. The denominator a is replaced by the double $2a$ and $4a^3 + 3a$. Both are significantly larger than a, thus increasing the speed of approximation. At the same time, the equation can be used to compare angles, whose tangent value are $1/a$ and $1/(2a)$. Here is a proof of the equation.

We use the same procedure as above and start with the triangle BC, where the tangent value of the angle α is $\tan(\alpha) = 1/(2a)$ (see Fig. 7.10). We enlarge this triangle by the factor $2a$ and on top of the construction we place a copy of the triangle ABC that is rotated by 90°. We see that the vertex of the construction $D = \left(4a^2 - 1, 4a\right)$ lies slightly above the line $y = x/a$, which has the tangent $1/a$. Now we find the correction angle HAD, moving vertically from the point $D = \left(4a^2 - 1, 4a\right)$ in the direction of the line $y = x/a$. Thus, the line DH has the following equation

$$\frac{y - 4a}{x - 4a^2 + 1} = -a.$$

To find the intersection point H with the line $y = x/a$, we set

$$\frac{x/a - 4a}{x - 4a^2 + 1} = -a$$

and find

$$x_H = \frac{4a^4 + 3a^2}{a^2 + 1} \text{ and } y_H = \frac{4a^3 + 3a}{a^2 + 1}.$$

Now we have

$$\tan(HAD) = \frac{DH}{AD} = \frac{4a - y_H}{x_H} = \frac{4a\left(a^2 + 1\right) - 4a^3 - 3a}{4a^4 + 3a^2} = \frac{4a - 3a}{4a^4 + 3a^2} = \frac{1}{4a^3 + 3a}.$$

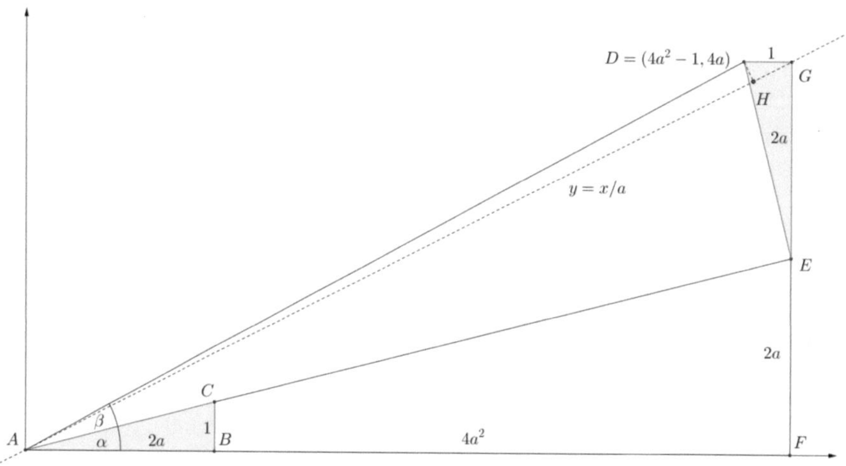

Fig. 7.10 Helpful arctangent identity

In total, we then have

$$\arctan\left(\frac{1}{a}\right) = 2\arctan\left(\frac{1}{2a}\right) - \arctan\left(\frac{1}{4a^3 + 3a}\right).$$

If we replace arctan(1/5) in Machin's formula with the help of our equation for $a = 5$, we obtain

$$\begin{aligned}\frac{\pi}{4} &= 4\arctan\left(\frac{1}{5}\right) - \arctan\left(\frac{1}{239}\right)\\ &= 8\arctan\left(\frac{1}{10}\right) - \arctan\left(\frac{1}{239}\right) - 4\arctan\left(\frac{1}{515}\right),\end{aligned}$$

a pretty good formula, compared to our first attempt for $a = 10$. This formula was discovered by the Scottish mathematician Robert Simson in 1723. However, this formula also includes the term arctan(1/239). This means that the error in the approximation of π is approximately $4 \cdot R_2 \approx 4 \cdot 2.442 \cdot 10^{-8} \approx 10^{-7}$ and we can expect at most six to seven correct decimal places after the decimal point, which is only a modest improvement compared to Machin's formula.

To understand the patterns that appear above to some extent, we now assume that n/a is already a good approximation to $\pi/4$, so $\pi \approx 4n/a$. Then we have

$$4 \cdot 2n\arctan\left(\frac{1}{2a}\right) - 4n\arctan\left(\frac{1}{a}\right) = 4n\arctan\left(\frac{1}{4a^3 + 3a}\right) < \frac{4n}{4a^3 + 3a} \approx \frac{\pi}{4a^2}.$$

As soon as $a \geq 219$, the value $\frac{\pi}{4a^2} < 1.6 \cdot 10^{-5}$, is quite small, and the approximation $4 \cdot 2n \arctan\left(\frac{1}{2a}\right)$ for π is very close to our starting approximation $4n \arctan\left(\frac{1}{a}\right)$. This applies to doublings of the denominator. The situation is similar for other multiples. If we again assume that $\pi \approx 4n/a$, then we have

$$\begin{aligned} &4 \cdot kn \arctan\left(\frac{1}{ka}\right) - 4n \arctan\left(\frac{1}{a}\right) \\ &= 4kn\left(\frac{1}{ka} - \frac{1}{3(ka)^3} + \cdots\right) - 4n\left(\frac{1}{a} - \frac{1}{3a^3} + \cdots\right) \approx \frac{4n}{3a^3} \approx \frac{\pi}{3a^2}. \end{aligned}$$

Thus, the multiples of $a = 219$ again yielded useful approximations, as the approximation for $a = 219$ itself was very good. However, this is not yet a sufficient explanation for the emerging pattern, especially the additional terms that represent multiples of 14. This leaves room for further research by the reader.

7.15 Third Approach

In a third approach, we now want to consider an even more general form of arctangent identities, where we gain two new degrees of freedom

$$\frac{\pi}{4} = n \cdot \arctan\left(\frac{1}{a}\right) + m \cdot \arctan\left(\frac{1}{b}\right) + \arctan\left(\frac{M}{N}\right).$$

In addition to the values of a and n, we can also vary the values of b and m, which gives us hope for new good approximations. A corresponding computer program will therefore contain two extra loops, one for b and one for m. Otherwise, the program looks very similar to the one above. We leave it to the reader to create such a program.

We first choose values for a and n and thus specify an angle $n \cdot \arctan(1/a)$. If this angle is less than 45°, we fill up with multiples of the angle $\arctan(1/b)$ until we reach the 45° mark again, by successively multiplying with the complex number $b+i$. Here we are again interested in those angles, which occur just before and just after crossing the 45° mark, very similar to Fig. 7.9. Because there we can again expect small residual angles and therefore also "good" arctangent identities.

Here are the results for $3 \leq b < a < 73$ and $\frac{M}{N} < 2 \cdot 10^{-5}$.

Unfortunately, no obvious pattern can be observed here. However, it is immediately apparent that the tangent values of the remaining angles all fall within the range of 10^{-6} with only very few exceptions that are significantly better. This gives us another reason to highlight the peculiarity of the case $a = 28$ and $n = 22$ which resulted in a remarkably small remaining angle.

The best value from Table 7.7 is $a = 71, n = 28, b = 23$ and $m = 9$, where the tangent of the remaining angle is $1.8 \cdot 10^{-8}$. Here, we have

$$\frac{\pi}{4} = 28 \cdot \arctan\left(\frac{1}{71}\right) + 9 \cdot \arctan\left(\frac{1}{23}\right) + \arctan\left(\frac{M}{N}\right),$$

where $M/N < 1.8 \cdot 10^{-8}$. If we, again, use four terms of the Gregory series for the calculation of $\arctan(\frac{1}{71})$ and $\arctan(\frac{1}{23})$, the associated errors amount to

$$|R_1| = \left|\frac{x^9}{9} - \frac{x^{11}}{11} + \cdots\right| \leq \left|\frac{x^9}{9}\right| = \frac{1}{71^9 \cdot 9} < 2.24 \cdot 10^{-18}$$

and

$$|R_1'| = \left|\frac{x^9}{9} - \frac{x^{11}}{11} + \cdots\right| \leq \left|\frac{x^9}{9}\right| = \frac{1}{23^9 \cdot 9} < 6.16 \cdot 10^{-14}$$

For the third arctangent value, we again use only a single term from the series. This results in

$$|R_2| = \left|\frac{x^3}{3} - \frac{x^5}{5} + \cdots\right| \leq \left|\frac{x^3}{3}\right| < \frac{1.8^3 \cdot 10^{-24}}{3} < 1.95 \cdot 10^{-24}.$$

All in all, the total error is then at most

$$\begin{aligned}&\left|4 \cdot 28 \cdot R_1 + 4 \cdot 9 \cdot R_1' + 4 \cdot R_2\right| \\ &< 4 \cdot 28 \cdot 2.24 \cdot 10^{-18} + 4 \cdot 9 \cdot 6.16 \cdot 10^{-14} + 4 \cdot 1.95 \cdot 10^{-24} < 10^{-11},\end{aligned}$$

and we could determine π to 10 decimal places accurately in this way, which is not exactly overwhelming. Once again, the situation $a = 28, n = 22$ and $M/N = 1.77 \cdot 10^{-5}$ proves to be a very special stroke of luck.

In another computer program, we assume that we have already exceeded the 45° mark with multiples of $\arctan(1/a)$. To get back to the vicinity of the line $y = x$, we strictly speaking have to divide by $b + i$, which leads to the subtraction of angles. This poses some difficulties for the computer because we are leaving the range of whole numbers. So far, our formulas, $x = ax - y$ and $y = x + ay$ have always given us whole numbers. A division would take us out of this number range and the computer would continue to calculate with decimal numbers, which in turn would cause new errors over which we have no control. Here, a small trick helps us.

A division of complex numbers is represented as follows:

$$\frac{p + qi}{r + si} = \frac{(p + qi)(r - si)}{(r + si)(r - si)} = \frac{(p + qi)(r - si)}{r^2 + s^2}.$$

Table 7.7 The results of the computer program from the third approach

a	n	b	m	M/N	a	n	b	m	M/N
30	15	28	8	$-9.67 \cdot 10^{-6}$	32	8	28	15	$-7.19 \cdot 10^{-6}$
35	5	28	18	$-4.15 \cdot 10^{-6}$	35	10	26	13	$7.91 \cdot 10^{-6}$
35	12	9	4	$5.38 \cdot 10^{-6}$	35	25	14	1	$7.4 \cdot 10^{-7}$
40	20	21	6	$3.67 \cdot 10^{-6}$	42	3	28	20	$8.3 \cdot 10^{-7}$
42	21	21	6	$6.00 \cdot 10^{-6}$	44	12	39	20	$9.66 \cdot 10^{-6}$
45	5	43	29	$8.27 \cdot 10^{-6}$	48	29	11	2	$8.9 \cdot 10^{-7}$
51	13	49	26	$9.73 \cdot 10^{-6}$	51	37	50	3	$8.91 \cdot 10^{-6}$
53	32	22	4	$3.09 \cdot 10^{-6}$	54	6	43	29	$2.68 \cdot 10^{-6}$
55	9	8	5	$4.85 \cdot 10^{-6}$	55	15	39	20	$7.23 \cdot 10^{-6}$
56	2	28	21	$6.30 \cdot 10^{-6}$	56	4	28	20	$5.07 \cdot 10^{-6}$
56	13	9	5	$6.12 \cdot 10^{-6}$	56	16	24	12	$3.29 \cdot 10^{-6}$
56	36	14	2	$5.59 \cdot 10^{-6}$	57	5	10	7	$7.29 \cdot 10^{-6}$
57	30	27	7	$4.44 \cdot 10^{-6}$	57	37	44	6	$1.77 \cdot 10^{-6}$
58	22	32	13	$7.56 \cdot 10^{-6}$	60	8	46	30	$5.96 \cdot 10^{-6}$
61	9	58	37	$2.56 \cdot 10^{-6}$	61	10	37	23	$8.08 \cdot 10^{-6}$
63	7	43	29	$6.7 \cdot 10^{-7}$	63	18	24	12	$3.07 \cdot 10^{-6}$
63	19	62	30	$7.17 \cdot 10^{-6}$	64	11	44	27	$6.40 \cdot 10^{-6}$
64	33	63	17	$3.49 \cdot 10^{-6}$	65	2	53	40	$3.91 \cdot 10^{-6}$
65	33	54	15	$1.2 \cdot 10^{-7}$	66	23	16	7	$8.30 \cdot 10^{-6}$
69	18	61	32	$3.68 \cdot 10^{-6}$	69	22	45	21	$9.96 \cdot 10^{-6}$
70	5	28	20	$7.80 \cdot 10^{-6}$	70	12	57	35	$9.15 \cdot 10^{-6}$
70	20	24	12	$7.63 \cdot 10^{-6}$	71	4	48	35	$2.65 \cdot 10^{-6}$
71	19	56	29	$8.11 \cdot 10^{-6}$	71	23	52	24	$5.65 \cdot 10^{-6}$
71	28	23	9	$1.8 \cdot 10^{-8}$	72	8	43	29	$2.86 \cdot 10^{-6}$
72	23	15	7	$2.88 \cdot 10^{-6}$					

That is, the division by $r + si$ can be described as a multiplication with the complex number $r - si$ and a subsequent division by the real number $r^2 + s^2$. However, the latter does not change the angle. But at the moment we are only interested in the subtraction of angles, while we are not interested in the length of the resulting complex numbers (vectors). Therefore, in our case, we can simply omit the division by $r^2 + s^2$ and arrive at a result with the same angle, see Fig. 7.11. At the same time, the associated calculations are again carried out in the range of whole numbers, and we do not lose any accuracy.

This means that if our angle $n \cdot \arctan(1/a)$ is above 45°, we can "bring it back" into the vicinity of 45° by repeated multiplication with the complex number $b - i$.

In Table 7.8 the results of this investigation are collected for $3 \leq b < a < 70$ and $|M/N| < 2 \cdot 10^{-5}$.

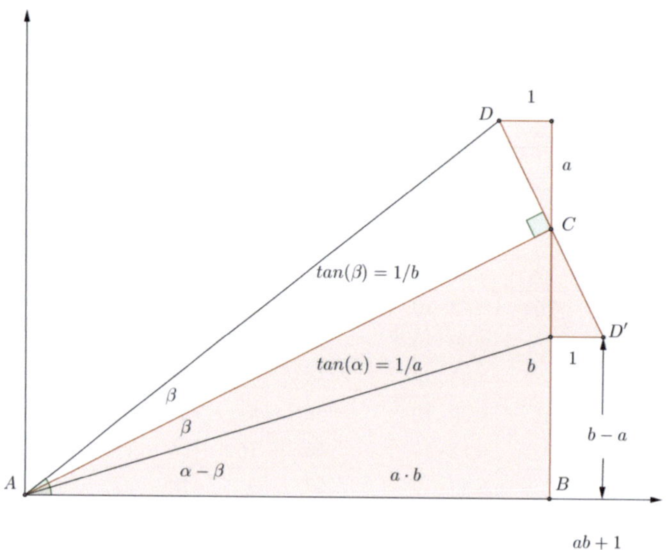

Fig. 7.11 Subtraction of angles

This investigation also shows that the tangent values of the remaining angles are approximately in the same range, around 10^{-5} and 10^{-6}. No significant improvements are observed. Unfortunately, no clear pattern can be discerned here either. Thus, our result from the previous program $a = 28, \quad n = 22$ and $M/N = 1.77 \cdot 10^{-5}$ remains as a special individual case.

7.16 Calculations of π Today

For some time now, mathematicians have been competing to calculate the most digits of π. Georg von Vega (1789), who calculated π to 126 digits accurately, and Zacharias Dahse, who calculated π to 200 digits accurately (1844), have already been mentioned. The competition continues, but today we are talking about 100 trillion digits, found with the help of computers. Completely different methods are used than those we have presented here.

The record set by Yasumasa Kanada from Tokyo University in 2002 was still achieved with the help of arctangent identities, similar to those in the Machin formula [Records], namely with the help of the two arctangent identities

$$\frac{\pi}{4} = 12 \arctan\left(\frac{1}{49}\right) + 32 \arctan\left(\frac{1}{57}\right) - 5 \arctan\left(\frac{1}{239}\right) + 12 \arctan\left(\frac{1}{110{,}443}\right)$$

Table 7.8 The results of the computer program with subtraction of angles

a	n	b	m	M/N	a	n	b	m	M/N
18	27	14	10	$13.1 \cdot 10^{-6}$	19	20	15	4	$9.5 \cdot 10^{-6}$
21	20	12	2	$18.5 \cdot 10^{-6}$	21	27	16	8	$4.85 \cdot 10^{-6}$
22	22	14	3	$8.41 \cdot 10^{-6}$	24	36	21	15	$11.8 \cdot 10^{-6}$
28	36	26	13	$16.2 \cdot 10^{-6}$	29	29	28	6	$11.0 \cdot 10^{-6}$
29	37	4	2	$-1.24 \cdot 10^{-6}$	31	28	17	2	$-2.89 \cdot 10^{-6}$
32	41	6	3	$11.0 \cdot 10^{-6}$	37	34	30	4	$-13.1 \cdot 10^{-6}$
39	33	33	2	$17.2 \cdot 10^{-6}$	39	46	33	13	$-11.7 \cdot 10^{-6}$
44	58	15	8	$-11.4 \cdot 10^{-6}$	45	44	26	5	$-5.78 \cdot 10^{-6}$
45	50	43	14	$-7.38 \cdot 10^{-6}$	47	42	37	4	$-2.25 \cdot 10^{-6}$
49	51	47	12	$6.93 \cdot 10^{-6}$	49	57	45	17	$9.92 \cdot 10^{-6}$
49	59	43	18	$12.8 \cdot 10^{-6}$	49	60	41	18	$15.6 \cdot 10^{-6}$
49	62	25	12	$11.9 \cdot 10^{-6}$	51	59	35	13	$11.1 \cdot 10^{-6}$
52	69	24	13	$-7.84 \cdot 10^{-6}$	53	52	46	9	$3.85 \cdot 10^{-6}$
53	80	29	21	$-5.59 \cdot 10^{-6}$	54	48	29		$18.1 \cdot 10^{-6}$
54	59	13	4	$18.0 \cdot 10^{-6}$	54	64	20	8	$15.5 \cdot 10^{-6}$
54	71	34	18	$-7.19 \cdot 10^{-6}$	55	74	50	28	$17.1 \cdot 10^{-6}$
55	85	50	38	$12.5 \cdot 10^{-6}$	57	74	39	20	$-6.11 \cdot 10^{-6}$
57	85	34	24	$1.97 \cdot 10^{-6}$	57	88	29	22	$17.1 \cdot 10^{-6}$
58	52	27	3	$-4.39 \cdot 10^{-6}$	58	52	36	4	$17.8 \cdot 10^{-6}$
59	51	38	3	$3.29 \cdot 10^{-6}$	59	66	24	8	$1.82 \cdot 10^{-6}$
59	84	47	30	$7.24 \cdot 10^{-6}$	60	64	32	9	$-11.2 \cdot 10^{-6}$
60	65	47	14	$-7.47 \cdot 10^{-6}$	61	70	58	21	$-6.94 \cdot 10^{-6}$
61	75	18	8	$-11.8 \cdot 10^{-6}$	61	94	45	34	$-16.2 \cdot 10^{-6}$
62	58	60	9	$-18.4 \cdot 10^{-6}$	62	63	26	6	$12.7 \cdot 10^{-6}$
62	65	19	5	$17.2 \cdot 10^{-6}$	62	66	43	12	$-6.19 \cdot 10^{-6}$
63	56	29	3	$-8.78 \cdot 10^{-6}$	63	82	62	32	$4.44 \cdot 10^{-6}$
64	52	37	1	$-15.2 \cdot 10^{-6}$	64	57	19	2	$11.7 \cdot 10^{-6}$
64	69	41	12	$-14.2 \cdot 10^{-6}$	64	89	38	23	$9.81 \cdot 10^{-6}$
64	97	63	46	$-6.09 \cdot 10^{-6}$	64	100	18	14	$4.37 \cdot 10^{-6}$
65	90	15	9	$5.47 \cdot 10^{-6}$	65	94	56	37	$2.46 \cdot 10^{-6}$
66	69	50	13	$-11.0 \cdot 10^{-6}$	66	81	43	19	$0.17 \cdot 10^{-6}$
66	83	36	17	$19.4 \cdot 10^{-6}$	66	94	36	23	$-10.6 \cdot 10^{-6}$
67	61	40	5	$-8.03 \cdot 10^{-6}$	67	61	48	6	$-0.08 \cdot 10^{-6}$
67	61	56	7	$4.71 \cdot 10^{-6}$	67	61	64	8	$7.82 \cdot 10^{-6}$
67	83	11	5	$-16.3 \cdot 10^{-6}$	67	94	34	21	$-13.6 \cdot 10^{-6}$
67	98	31	21	$4.72 \cdot 10^{-6}$	68	66	27	5	$-19.5 \cdot 10^{-6}$

(continued)

Table 7.8 (continued)

a	n	b	m	M/N	a	n	b	m	M/N
68	67	20	4	$8.64 \cdot 10^{-6}$	68	77	49	17	$17.4 \cdot 10^{-6}$
68	92	37	21	$-16.0 \cdot 10^{-6}$	68	98	61	40	$4.54 \cdot 10^{-6}$
69	85	56	25	$-18.5 \cdot 10^{-6}$	69	87	61	29	$-15.8 \cdot 10^{-6}$
69	93	32	18	$-16.5 \cdot 10^{-6}$	69	101	28	19	$15.6 \cdot 10^{-6}$

by K. Takano (1982) and

$$\frac{\pi}{4} = 44\arctan\left(\frac{1}{57}\right) + 7\arctan\left(\frac{1}{239}\right) - 12\arctan\left(\frac{1}{682}\right) + 24\arctan\left(\frac{1}{12{,}943}\right)$$

by Carl Størmer (1896). In this process, 1,241,100,000,000, or over a trillion digits, of π were calculated. Since then, however, nearly all records have been achieved using other methods based on the series of Srinivasa Ramanujan, who published several extremely rapidly converging series in 1910. One of them looks like this:

$$\frac{1}{\pi} = \frac{2\sqrt{2}}{9{,}801}\sum_{k=0}^{\infty}\frac{(4k)!(1{,}103 + 26{,}390k)}{(k!)^4 396^{4k}}.$$

This is a so-called hypergeometric series, i.e., a geometric series with a linear term in the numerator, where an additional factor of the form $(4k)!/(k!)^4$ is incorporated. Now we have

$$\frac{(4k)!}{(k!)^4} = \frac{1\cdot 2\cdot 3\cdot 4}{1\cdot 1\cdot 1\cdot 1}\cdot\frac{5\cdot 6\cdot 7\cdot 8}{2\cdot 2\cdot 2\cdot 2}\cdots\frac{(4k-3)\cdot(4k-2)\cdot(4k-1)\cdot 4k}{k\cdot k\cdot k\cdot k} < 4^{4k}.$$

Thus, we see that the series converges. It even converges very rapidly. It produces at least 8 new correct decimal places for π for each new term. In 1989, the Chudnovsky brothers calculated a billion decimal places of π, using a variant of the Ramanujan series

$$\frac{1}{\pi} = 12\sum_{k=0}^{\infty}\frac{(-1)^k(6k)!(13{,}501{,}409 + 545{,}140{,}134k)}{(3k)!(k!)^3 640{,}320^{3k+1/2}}$$

Since then, most records have been achieved with the help of the Chudnovsky algorithm. The latest record was set by Emma Haruka Iwao on June 8, 2022. With the help of the Google Cloud Blog, she was able to calculate 10^{14} places of π. The calculations took 158 days [Record 2022]. This record was tied in 2023, when

Jordan Ranous also calculated 10^{14} places of π. However, he managed this in only 59 days (54 days) [Record 2023 A and Record 2023 B].

The mathematics behind the Ramanujan series and the Chudnovsky algorithm would go beyond the scope of this book. We also see that the mentioned series contain root functions, so we have to leave the area of integers and fractions and also have to use dedicated rapidly converging algorithms for the calculation of the roots. Of course, there are also fast algorithms for these tasks, but the problem becomes increasingly complex and the control over possible errors increasingly eludes us.

Remark 7.4

Here we want to show how to explain the multiplication of complex numbers (not just those of the form $a + i$) using the above geometric construction. We multiply $a + bi$ by $c + di$.

We start with Fig. 7.12, where the first complex number $a+bi$ is represented by the hypotenuse in the right-angled triangle ABC with $\tan(CAB) = \frac{b}{a}$. Now we enlarge the construction by the factor c and obtain the triangle $AB'C'$, where $C' = (ac, bc)$. On this point, we set an enlarged version of the original triangle ABC, where we have used the enlargement factor d and rotated the triangle by 90°. Then the vertex D of our construction has the coordinates $D = (ac - bd, bc + ad)$ and we see the connection to complex multiplication. We have $\tan(CAD) = \frac{C'D}{AC'} = \frac{d}{c}$, and therefore the angle BAD is just the sum of the angles for $a + bi$ and $c + di$.

As for the length of the line AD, i.e., the magnitude of the product, we set $l = \sqrt{a^2 + b^2}$ and $h = \sqrt{c^2 + d^2}$ for the lengths of the two complex numbers. Because of the enlargements then $AC' = cl$ and $C'D = dh$. Then it follows

$$AD^2 = AC'^2 + C'D^2 = c^2l^2 + d^2l^2 = l^2(c^2 + d^2) = l^2h^2.$$

This is now the square of the product of the lengths of the two complex numbers that we had multiplied. This shows that in the present geometric representation of the multiplication of complex numbers, the lengths of the factors multiply to the length of the product, while the angles add to the angle of the product.

7.17 Summary

The classical methods for calculating π, as found by Archimedes and Zu Chongzhi, are based on doubling the sides of regular polygons. From a computational point of view, root calculations occur here, whereby the calculations then leave the realm of rational numbers. In order to be able to study rational approximations to π nevertheless, in this chapter we turn to a different approach that was developed in the seventeenth century. This is based on two important ingredients. Firstly, we look at arctangent identities and secondly, we use the power series

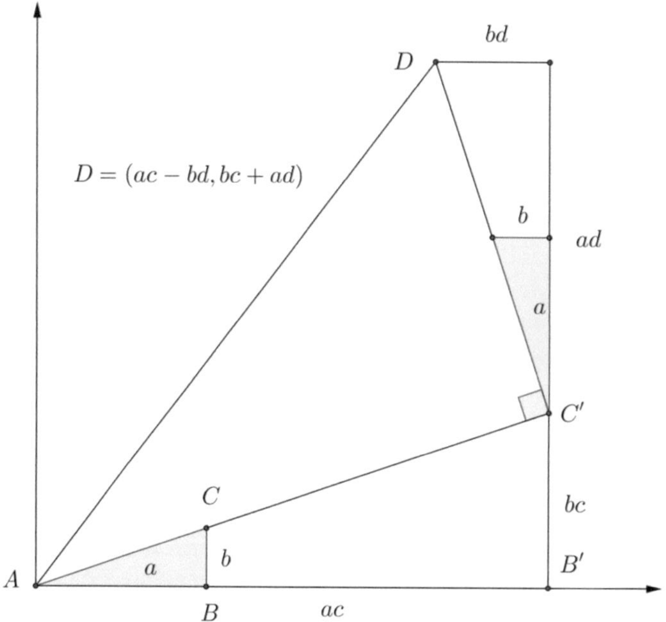

Fig. 7.12 Multiplication of complex numbers

expansion of the arctangent function. It turns out that arctangent identities can be generated in many different ways, depending on how many angles you want to include. This opens up a field where the interested reader can also carry out their own investigations and possibly find exciting results. Above all, further research can be carried out with the aid of computers and - as we show in the course of the chapter - astonishing patterns can be discovered in the "good" approximations. So, there is still some residual potential inherent in this method that is yet to be uncovered. At the end of the chapter, however, we also see that the method with the arctangent identities is no longer used by today's π-hunters. More effective approaches are now at the center of investigations and, above all, in the creation of records.

Bibliography

Nelsen. R., B. (2016) *Proofs Without Words III, Further Excercises in Visual Thinking*, Mathematical Association of America.

Internet Addresses

[Approximations to pi] https://en.wikipedia.org/wiki/Approximations_of_%CF%80

[5000 terms] https://en.wikipedia.org/wiki/Leibniz_formula_for_%CF%80
[Hutton] https://mathoverflow.net/questions/163242/geometric-explanation-of-huttons-formula
[Machin 1] https://mathworld.wolfram.com/Machin-LikeFormulas.html
[Machin 2] https://en.wikipedia.org/wiki/John_Machin
[Milü] https://en.wikipedia.org/wiki/Mil%C3%BC
[Records] https://en.wikipedia.org/wiki/Chronology_of_computation_of_%CF%80#The_age_of_electronic_computers_(from_1949_onwards)
[Record 2022] "*Even more pi in the sky: Calculating 100 trillion digits of pi on Google Cloud*". Google Cloud Platform.
[Record 2023 A] https://www.storagereview.com/review/storagereview-calculated-100-trillion-digits-of-pi-in-54-days-besting-google-cloud
[Record 2023 B] https://news.solidigm.com/en-WW/225029-solidigm-helps-break-world-record-pi-calculation
[Strassnitzky] https://mathstats.uncg.edu/sites/math-bio-fellowship/preprints/GoRy07b.pdf
[Zu Chongzhi] https://en.wikipedia.org/wiki/Mil%C3%BC

MIX
Papier aus verantwortungsvollen Quellen
Paper from responsible sources
FSC® C105338

If you have any concerns about our products,
you can contact us on
ProductSafety@springernature.com

In case Publisher is established outside the EU,
the EU authorized representative is:
Springer Nature Customer Service Center GmbH
Europaplatz 3, 69115 Heidelberg, Germany

Printed by Libri Plureos GmbH
in Hamburg, Germany